A Primer for Model-Based Systems Engineering

2nd Edition

David Long
and
Zane Scott

Copyright © 2011 Vitech Corporation. All rights reserved.

ISBN 978-1-105-58810-5

Permission to reproduce and use this document or parts thereof and to prepare derivative works from this document is granted, provided that both attribution to Vitech Corporation and this copyright notice are included with all reproductions and derivative works.

Product names mentioned herein are used for identification purposes only, and may be trademarks of their respective companies.

Publication Date: October 2011

DEDICATION

We dedicate this edition of the Vitech MBSE Primer to the memory of our friend and mentor, Jim Long. He blazed the trail in this discipline from his days as a TRW engineer to his years as our Chief Methodologist. His stories of his many experiences along the way formed a light bright enough to illuminate the path forward for us and the many others whose lives he touched. He always encouraged us to be more than we thought possible. In a letter to Robert Hooke, Isaac Newton wrote, "If I have seen further it is only by standing on the shoulders of giants." Jim is surely the giant on whose shoulders we have stood to see the way forward. This is for him.

Vitech Corporation Research and Education Council

CONTENTS

INTRODUCTION i
THE PROBLEM: A GEOSPATIAL LIBRARY v
WHAT IS A SYSTEM? 1
    SYSTEMS THINKING 4
    APPLYING SYSTEMS THINKING TO SYSTEMS DESIGN 7
WHAT IS SYSTEMS ENGINEERING? 11
    MULTIDISCIPLINARY APPROACH 13
    PROBLEM CLASSES 13
    THE DESIGN SPACE: THREE SYSTEMS 16
    THE DESIGN SPACE: BOUNDARIES 20
    THE PROCESS 21
    DOMAINS 24
    COMMUNICATION 29
WHAT IS A MODEL? 31
    FOUR ELEMENTS OF A MODEL 32
    CHARACTERISTICS OF A MODEL 33
    LANGUAGE: THE SYSTEMS MODEL IS LANGUAGE-BASED 34
    LANGUAGE OF BEHAVIOR 39
    MANAGING COMPLEXITY WITH LANGUAGE 42
    STRUCTURE: THE MODEL EXPRESSES SYSTEM RELATIONSHIPS 44
    ARGUMENT: THE MODEL IS USED TO "PROVE" THE CONCEPT OF THE DESIGN ... 53
    PRESENTATION: THE MODEL MUST BE "VIEWABLE" 54
WHAT IS MODEL-BASED SYSTEMS ENGINEERING? 65
    REQUIREMENTS FOR A SYSTEMS ENGINEERING PROCESS 68
    MBSE MODEL AND SYSTEM DEFINITION LANGUAGE 74
    DEVELOPING LAYER 1 OF OUR SOLUTION 76
    PROCEEDING WITH LAYER 2 85
    ARCHITECTURE DESIGN AT LAYER N 95
    VERIFICATION AND VALIDATION 97
SUMMARY 103
AFTERWORD 105
AUTHORS 107

A Primer for Model-Based Systems Engineering

INTRODUCTION

This is the 2nd edition of Vitech's model-based systems engineering primer. In this second treatment of the subject, we have covered the same subject matter as before but augmented this time with what we have learned since releasing the 1st edition. We strive to be a "learning organization" and to leverage that learning for the benefit of our customers and community. With this edition we hope to carry that principle forward.

There are notable differences in this edition. First, we have reorganized the material. Instead of the topical organization of the 1st edition, we have approached the description of model-based systems engineering (MBSE) from a "building blocks" perspective suggested by its name. We ask first "What Is a System?" From there we tackle "What Is Systems Engineering?" Then we discuss "What Is a Model?" and finally arrive at the question "What Is Model-Based Systems Engineering?" We hope that this building approach will make it easier to put the concepts into a logical framework for understanding and use.

We have also tied the concept discussions more closely to practical illustrations. We have largely drawn these from the example system design included with this primer. This has been done in response to many helpful suggestions from our readers, and we think it makes the concepts much clearer and easily understood.

One of the most common flaws in any undertaking is a departure from the fundamental principles of the disciplines involved in the process. This can be due to inattention bred by familiarity or a failure to recognize and reinforce "the basics." Whether the enterprise is a football game or a systems design project, the fundamentals of "blocking and tackling" are critical to success. Absent or poorly executed, they can doom the venture. In the case of a floundering effort, they are the key to getting back on track.

The importance of knowing and executing the basics is the driving force behind this primer. It is the reason for not beginning with a collection of essays on more advanced topics. Revisiting the "blocking and tackling" aspects of MBSE is the foundation of our effort to advance the cause of sound systems design.

This primer addresses the basic concepts of model-based systems engineering. It covers the Model, Language, Behavior, Process, Architecture, and Verification and Validation. It is a call to consider the foundational principles behind those concepts. It is not designed to present novel insights into MBSE so much as to provide a guided tour of the touchstones of systems design. It is a guide to the new MBSE acolyte and a reminder to the experienced practitioner.

Why such a basic approach? Without this grounding, it can become easy to lose the sense of relationship between techniques and the design itself. Reading and pondering the sections on Models and Language bring into focus the difference between representations of the model and the model itself. A map may be an extensive, informative, and important representation of the underlying terrain, but it is simply that—a representation. Likewise, a set of diagrams may be useful, clear, and detailed, but they are not the model of the system itself. Without returning to the concepts of model, language, process, and behavior, we can easily become mistakenly convinced that the process of drafting a "full set" of representations is the same thing as constructing a model. It is through understanding the basics that we understand the distinctions.

A Primer for Model-Based Systems Engineering

In other ways as well, the failure to be aware and alert to the basic principles of MBSE can hinder the integrity of a system-design effort. Just as a football team returns again and again to their roots in basic skills, we can all profit from reacquainting ourselves with these basic principles.

This primer is offered to the end that it will function as a call back to the basic concepts of our discipline. It lays the groundwork for improvements and enhancements already being planned. As it stands, it is a look at the foundational concepts of MBSE designed to benefit the newcomer and experienced practitioner alike.

Finally, for whom is this primer intended? Of course the obvious answer is that it is intended to be an introduction to these concepts for those who may be new to the world of model-based systems engineering. It is written in a way that can be understood by any intelligent and curious reader—even if that reader is not an engineer. Project managers, acquisitions professionals, and business-process consultants can all use this primer to guide them into the MBSE concepts in an organized way.

In addition, this primer is intended to provide an organized presentation of these concepts for the systems engineering practitioner who may need a reference framework for them. Often we become familiar with the concepts we use in the way that we customarily use them. This is as true in the systems engineering discipline as in any other. Like a gradually fragmenting hard drive, our thinking becomes compartmentalized in ways that reflect how much and when we use the concepts we have learned over time.

With this primer we hope to provide the seasoned practitioner a framework in which to refresh the concepts and see the relationships. Any experienced systems engineer can take these and expand them with additional detail and application anecdotes. It is our hope that this basic discussion can provide a "rack" in which to place that experience and, thereby, make it more useful in practice.

It is our hope that you find this primer valuable. We welcome your comments and suggestions about improving it. Much of what we have learned about how it should be organized and presented has come from thoughtful contributions from the readers of the 1st edition.

Vitech Corporation

October 2011

THE PROBLEM: A GEOSPATIAL LIBRARY

Throughout this primer, we will consider examples drawn from the following sample system design problem. In accord with our ultimate destination, a layered approach to model-based systems engineering (MBSE), we will begin our description of the system at a very high level.

In this example, we present the need for the system as a discrete design problem. The underlying need is for a system that will allow a set of image collectors (in our case, satellites) to collect images for a set of customers and provide those images to the customers. This high-level system description could be represented as follows:

Geospatial Library

CUSTOMER   MANAGEMENT SYSTEM   COLLECTOR

This drawing depicts the customer making a request of the management system for an image to be produced by the collector. The management system tasks the collector to produce the image, the collector gathers it and sends it back to the management system, whereupon it is provided back to the customer. The customer, the management system, and the collector are all shown separately.

In this problem, the system to be designed is the management system. It will interface with the customer and the collector, but the design and function of those systems are outside the scope of the problem. They simply impose conditions which their respective interfaces must meet.

It is obvious that the customer in this system must be able to make a request (place an order) for an image. The collector network, however, is a sensitive set of government satellites. The images produced can't be distributed to just anyone who decides to request them. A customer qualification process is needed to certify the eligibility to make requests.

These considerations are initially presented to the design team in a concept of operations (CONOPS) document. The system documentation is later expanded to include a source document which incorporates an engineering standards document by reference.

As the designers further explore the capability to accept customer requests, it becomes clear that customers need to be able to place orders in person, on the phone, by hardcopies delivered by messengers, by fax, and over the web. The capability to accommodate all five of these formats elaborates on the highest level requirement that the system "accept information requests from certified customers."

As the system design discussions unfold, it becomes apparent to the system stakeholders that a great deal of inefficiency will result if the system is unable to catalogue and store images that are taken. If another customer wants the same image, unless there is a searchable image library, the collector(s) will have to be retasked, and the work associated with gathering and processing that image will be duplicated. Therefore, the stakeholder has asked that the design include an image library capability.

There are also performance and resource limitations to be considered. The system will have only 25 people per shift to operate all functions. Performance standards will also include responding to customer requests within 24 hours. These standards are set out in the source document as requirements. In addition, the source document requires that the system must be available around the clock every day of the year. Availability is defined as having 10 minutes or less down time each month.

We will use this system design as a source of examples and discussions throughout the primer. From it we will illustrate the basic concepts of model-based systems design.

WHAT IS A SYSTEM?

Although the term *system* is defined in a variety of ways in the systems engineering community, most definitions are similar to the one used in the U.S. Department of Defense Architecture Framework (DoDAF)—"any organized assembly of resources and procedures united and regulated by interaction or interdependence to accomplish a set of specific functions." (*Department of Defense Dictionary of Military and Associated Terms*, its.bldrdoc.gov/fs-1037/dir-036/_5255.htm).

In her book *Thinking in Systems: A Primer* (Chelsea Green Publishing, 2008), Donella Meadows puts it somewhat more succinctly by saying "A system is an interconnected set of elements that is coherently organized in a way that achieves something." She goes on to point out that "A system is more than the sum of its parts. It may exhibit adaptive, dynamic, goal-seeking, self-preserving, and sometimes evolutionary behavior."

The idea that a system is "more than the sum of its parts" is picked up in the International Council on Systems Engineering (INCOSE) definition of a system. INCOSE defines a system as "a construct or collection of different entities that together produce results not obtainable by the entities alone" ("A Consensus of the INCOSE Fellows", www.incose.org/practice/fellowsconsensus.aspx).

There are some clear commonalities among these three definitions. First, any system must be made up of what Meadows refers to as "elements" (called an "assembly of resources and procedures" in the DoDAF definition and "entities" by INCOSE). These are the parts of the system that together form the whole.

In addition, a system must tie the parts together with relationships. What Meadows calls "interconnect(ion)" and "coherently organized" is for INCOSE a "construct" and for DoDAF an "organized assembly . . . united and regulated by interaction or interdependence." The concept of relationships is the second common characteristic of these definitions.

Finally, the system must have a purpose for which the elements are assembled. DoDAF calls this organizing purpose "to accomplish a set of specific functions." INCOSE sees it as the ability to "produce results not obtainable by the entities alone." For Meadows the system is "organized in a way that achieves something." In each definition the elements of the system are related to each other in ways that promote the accomplishment of a specific purpose that is beyond the capability of any of the parts acting alone.

It should be noted at this point that these aspects of a system are often construed narrowly in practice, causing our view of systems to be constrained or limited. In reality, systems exist wherever these three are present: parts, relationships, and a purpose. This primer will be intentionally broad in its view of where those aspects are present. This will enable us to see systems in places where heretofore we might not have expected them.

Some examples of systems include:

- A set of things working together as parts of a mechanism or an interconnecting network.
- A set of organs in the body with a common structure or function.
- A group of related hardware units or software programs or both, especially when dedicated to a single application.
- A major range of strata that corresponds to a period in time, subdivided into series.

- A group of celestial objects connected by their mutual attractive forces, especially moving in orbits about a center.

The entities or elements of a system constructed by humans can include people, hardware, software, facilities, policies, and documents. For any given system, this list is limited only by the set of things required to produce its system-level results. These results include system-level qualities, properties, characteristics, functions, behavior, and performance. The value added by the system as a whole, beyond that contributed independently by the parts, is primarily created by the relationships among the parts. In other words, the "value-add" of the system emerges in the synergy created when the parts come together.

The sample problem included with this primer is that of a Geospatial Library tasked with connecting a set of satellite imaging collectors with its customers. It is composed of parts. At the highest level, it has a Command Center Subsystem that manages the collection, storage, and retrieval of imagery products as well as a Workstation Subsystem that manages the translation of various incoming imagery requests into an internal common imagery collection request.

The system parts have relationships that define their interaction and the system's function. For example, when a customer request cannot be serviced from the system's current inventory, the Command Center Subsystem makes a specific collection request to a specific sensor to satisfy the customer's need. The products generated from this request are collected, added to the Geospatial Library inventory, and combined into a package for shipment to the customer by the Workstation Subsystem. Together, the parts of the system have allowed the system to take a specific customer request, obtain the images necessary to satisfy that request, and provide the images to the customer in response.

This fulfills the system's purpose—servicing the needs of the customers and collectors in facilitating the exchange of requests for images and the images themselves. At a very high level this illustrates the presence of the parts, relationships, and purpose in the sample system.

Systems Thinking

In order to create systems, it is necessary to engage in "systems thinking." Peter Senge's book *The Fifth Discipline* (Doubleday/Currency, 1990) introduced systems thinking into popular culture. However, it remains largely unappreciated and is honored mainly in the breach rather than the observance. Part of this is because systems thinking is practiced using too narrow a definition of "systems," and this narrowness limits the practice of systems thinking.

For a broad understanding of what is meant by "systems thinking," we will turn to one of the preeminent systems thinkers, Russell Ackoff. Notice his use of the three aspects of the systems definition (parts, relationships, and purpose) in defining systems thinking: "systems thinking looks at relationships (rather than unrelated objects), connectedness, process (rather than structure), the whole (rather than just its parts), the patterns (rather than the contents) of a system, and context" (R. Ackoff with H. Addison and A. Carey, *Systems Thinking for Curious Managers*, Triarchy Press, 2010).

Ackoff goes on to state, "Thinking systemically also requires several shifts in perception, which lead in turn to different ways to teach, and different ways to organize society." This statement is significant in two ways. First, Ackoff is observing that the move to systems thinking requires changing the way we think. In addition, he is showing that he sees systems (and systems thinking) quite broadly.

Taking his latter suggestion first, Ackoff is interested in the application of systems thinking beyond the classic boundaries of systems engineering. Coming from a business and process orientation (as opposed to an engineering orientation), Ackoff sees the concepts of systems and systems thinking as broadly applicable to business and even social process design. In his book *Redesigning Society* (R. Ackoff and S. Rovin, Stanford Business Books, 2003), he focuses on the systems aspects of public policy decision making. He is truly committed to the idea of seeing systems wherever the three aspects are present.

Perhaps his most important insight has to do with the "shifts in perception" or changes in thinking involved in thinking systemically. A major shift in thinking comes from moving away from the exclusively analytic approach that has characterized our thinking since the Enlightenment. This analytic approach, according to Ackoff, "is a three-step process: (1) take the thing or event to be understood apart; (2) explain the behavior or properties of the parts taken separately; and (3) aggregate the explanations of the parts into an understanding of the whole, the thing to be explained" (Ackoff and Rovin, Redesigning Society). Such "analytic thinking" takes our focus off of the system and orients it to the parts individually. This analytic, parts-oriented approach leads too often to ill-fated attempts to improve system performance by improving the parts of the system. Not only are such attempts typically fruitless, but they can actually damage overall system performance or even destroy the system.

What is needed is a different way of thinking, a way of approaching problems from a systems perspective. Ackoff calls this new approach "synthetic thinking." According to Ackoff, "Synthetic thinking is also a three-step process, each the opposite of the corresponding step of analysis: (1) identify one or more systems that contain the system to be explained; (2) explain the behavior of the containing system (or systems); and (3) disaggregate the understanding of the containing system into the role or function of the system to be explained" (Ackoff and Rovin, Redesigning Society). The critical idea here is that we begin not from a decomposition of the system into its parts but from the point of view of the system in its context.

In the book *Systems Thinking for Curious Managers*, Ackoff points out that "Managers should never accept the output of a technologically-based support system unless they understand exactly what the system does and why. Many managers who are unwilling to accept advice or support from subordinates whose activities they do not fully understand, are nevertheless willing to accept support from computer-based systems of whose operations they are completely ignorant. Management information systems are usually designed by technologists who understand neither management nor the difference between data and information. Combine such ignorance with a management that does not understand the system the technologists have designed, and one has a recipe for disaster or, if lucky, large expenditures that bring no return" (Ackoff, Russell; Addison, Herbert; Carey, Andrew; Gharajedaghi, Jamshid (2010). *Systems Thinking for Curious Managers: With 40 New Management f-Laws*).

The point here is that systems must be understood in the context of what they can do and the world in which they will do it. It is not enough to see the system as a sum of the operations of the component functions. It must been seen as a functioning whole. This is the systems viewpoint.

This viewpoint allows us to engage the system without losing sight of the context and purpose of the system as a whole. Effective systems thinking combines analytic and synthetic thinking. It is common to see analytic thinking without its synthetic sibling. Too often this results in the loss of systems perspective. At its worst, this becomes component engineering.

The loss of the systems perspective can be quite costly. When the consequences of a limited or missing systems view emerge during the design process—as when different design paths result in mutually exclusive constraints—the penalty is expensive rework. Cost and schedule suffer together as the system is reengineered to correct the problems.

Sometimes the missing perspective doesn't levy its price until the system is built. This is the failure that Ackoff calls out—the failure to "understand exactly what the system does and why." This leads easily to unintended consequences as the system interacts with its environment in unanticipated and unhelpful ways.

Applying Systems Thinking to Systems Design

A system begins with an idea that must be translated into reality. The theoretical idea of a system must link to the engineered system "reality" and vice versa (bidirectional linkage). The designers must also find a way to clearly show when and how the theory explains reality and how reality confirms their theory.

The system design must take into account the system properties. Within the boundary of a system, there are three kinds of properties:

Entities—These are the parts (things or substances) that make up a system. These parts may be atoms or molecules; larger bodies of matter like sand grains, raindrops, plants, or animals; or even components like motors, planes, missiles, etc.

Attributes—Attributes are characteristics of the entities that may be perceived and measured such as quantity, size, color, volume, temperature, reliability, maintainability, and mass.

Relationships—Relationships are the associations that occur between entities and attributes. These associations are based on cause and effect.

In order to explain the design, the engineers must use some form of expression. When taken together, the properties of the system—the entities, attributes, and relationships—form a system "language." This language is fundamental to being able to describe and communicate the system among the engineering team as well as to other stakeholders.

Using this language, the system can be represented hierarchically, allowing it to be understood as decomposable into meaningful subunits. These subunits are conventionally named:

- a *system* is composed of subsystems;
- *subsystems* in turn are composed of assemblies;
- *assemblies* are composed of subassemblies, and
- *subassemblies* are composed of parts.

It is important to note that what may be considered a "part" in the context of a particular system may be a complete "system" in its own right. This all depends upon the point from which the system is viewed and the resulting system boundary decisions.

Often the terms used in describing this hierarchy are not well specified; some engineers use the term *sub-subsystem*, others use the terms *component* and *subcomponent* in the hierarchy. Variant usage only contributes to confusion. In order to avoid such usage confusion, the term *component* is used here as an abstract term representing the physical or logical entity that performs a specific function or functions.

The parts of a system interact to produce the performance of the whole system. It is intuitively obvious that all parts of the system must be functioning as designed in order for the system to function properly. What is not so obvious is that improving the function of one of the parts, be that a subsystem or component or whatever it may be labeled, will not necessarily improve the functioning of the whole. This is because of the effects of interaction within the system. For example, improving the resolution of the images gathered by the collectors in the sample problem will not improve the product for the customer if the image inventory cannot process and deliver them. Any improvement must be considered from a perspective that looks across the system as a whole.

The system results at the customer level depend upon the performance of the entire system. While the components must be understood from the perspective of whether or not they can perform the behavior allocated to them by the system design, it is ultimately the performance of the system that matters. This must account not only for the capability to meet the needs of the stakeholders that drove the system creation but also for any extraneous consequences of system performance, particularly unintended or unplanned consequences. Understanding and practicing this is the very foundation of systems thinking.

WHAT IS SYSTEMS ENGINEERING?

According to INCOSE the responsibility of systems engineering is "creating and executing an interdisciplinary process to ensure that the customer and stakeholder's needs are satisfied in a high quality, trustworthy, cost efficient and schedule compliant manner throughout a system's entire life cycle" ("A Consensus of the INCOSE Fellows," http://incose.org/practice/fellowsconsensus.aspx). Boiled down to its essence, this means that the systems engineer is required to create and maintain a system that meets the customers' needs. That can be accomplished only when the focus of the systems engineer is on the whole system and the system's external interfaces.

Systems engineering is concerned with the design, building, and use of systems composed of concrete entities such as engines, machines, and structures. It is equally concerned with business systems, which are composed of processes. Engaging in systems engineering requires an organized means of thinking about those systems in their operational contexts. This way of thinking is the heart of systems engineering.

Systems engineering begins by identifying the needs of the users and the stakeholders to assure that the right problem is being addressed by the system. The systems engineer crafts those needs into a definition of the system, identifies the functions that meet those needs, allocates those functions to the system entities (components) and finally confirms that the system performs as designed and satisfies the needs of the user.

This is both a technical and a management process. The technical process addresses the analytical, design, and implementation efforts necessary to transform the operational need into a system of the proper size and configuration for its purpose. Along the way, it produces the documentation necessary to implement, operate, and maintain the system.

The management process supports the technical process by planning, assessing risks, integrating the various engineering specialties and design groups, maintaining configuration control, and continuously auditing the effort to ensure that cost, schedule, and technical performance objectives are satisfied. Together, the management and technical processes create the systems that will meet the customers' needs.

To be effective in all these areas, systems engineering must, therefore, provide an organized, repeatable, iterative, and convergent approach to developing complex systems. The approach must be "organized," because without an organized approach the details of the system under development will be overlooked, confused, and misunderstood. The approach must be "repeatable" so that it will apply to other system development efforts in a way that creates reasonable assurances of success. It should be both iterative and convergent, which means the engineering processes repeat at each level of system design and ensure the convergence of the development process to a solution.

The success of the development process rests on the ability of the systems engineer to maintain a system focus. We talked about systems thinking in the discussion of the essential characteristics of a system. The systems engineer must keep the vision of the entire system in mind while moving through the process of designing the system that will form the solution to meet the needs of the stakeholders. Losing this focus will cause the design to fail to meet those needs in one or perhaps many ways.

Multidisciplinary Approach

The discipline of systems engineering brings together branches of engineering and science in planning and developing solutions for the stakeholders' needs. By adopting a systems view of the problem and the possible solutions, systems engineers can draw on the different disciplines to design a solution that most effectively meets the needs of the stakeholders. The power of systems engineering comes from using this multidisciplinary approach to problem solving to satisfy the needs of stakeholders through creating or improving a system.

Every approach has its advantages and problems. While the multidisciplinary nature of the systems engineering team leverages the differing experience and expertise of the various disciplines, it also creates potential problems resulting from the variety of specialized vocabulary and ways of communicating that are customary in those disciplines. It is the job of the systems engineer to provide the coordination and communication that will allow the power of a multidisciplinary approach to benefit the problem-solving effort without being impeded by the potential miscommunication and friction between the disciplines.

Problem Classes

Systems engineering can be applied to three classes of problems: top-down or "clean-sheet" problems, middle-out or system-improvement problems, and reverse-engineering or system-replacement problems. The classic problems are the top-down designs. Often the other two problem classes are not even considered in discussions of systems engineering.

However, all three are significant and can benefit from the discipline and rigor of systems engineering.

Top-down engineering problems are the best known among the three problem classes. In these situations the engineering team is faced with designing an unprecedented system solution to stakeholder problems. These designs are also commonly called "greenfield" or "clean-sheet" development efforts. Such systems have many unknowns. Solving these unknowns often involves doing research, developing new materials, new components, and new manufacturing methods to provide all that is necessary to implement the solution.

The initial definition of a top-down problem is usually to be found in a document or set of documents setting out the high-level requirements for the system. The design process begins with an analysis of these requirements. From this analysis emerges a high-level description of the system which is then used in designing the solution system.

Experience is now suggesting that the top-down problems that have heretofore held center stage are becoming more the exception than the rule. This is due to the increasing complexity of our world of interconnected systems and technology. It is becoming rare that stakeholders truly have the freedom to design a system in isolation, creating a completely new solution. Most often the new solution must incorporate or interface with existing technology/systems, and these legacy components or interfaces must be accounted for in the new design. Confronting this problem class is known as "middle-out" engineering.

Middle-out engineering begins with modeling the "as is" state of all or a portion of a system. This provides an understanding of the existing "solution" and the supporting processes and technology. From there the engineer can begin to see what can and cannot be changed and where the opportunities for improvement or for meeting new requirements lie. This is the platform for designing an improved or "to be" set of solutions to customer-identified problems using the derived system-level requirements and the customer-developed problem statement(s).

This approach has been quite successful in process improvement and system-of-system settings. It is becoming clear that this problem class will be more and more the subject of system design needs. Systems engineers will need to become comfortable working in this arena in order to meet the needs of their customers.

Bottom-up or reverse engineering applies to upgrading or replacing legacy systems. Legacy systems may have been in operation for many years. They may have had extensive enhancements and fixes added over the life of the system. These are usually implemented without adequate documentation. If documentation exists, it typically contains a variety of omissions and errors.

The focus of the reverse engineering effort is to recover the original system-level requirements of the system as built and modified. Once the system-level requirements are recovered, these requirements and the newly specified requirements that drive the redesign are used to design and implement a replacement system that will offer both new and existing capabilities on a more sustainable platform.

The complete systems engineering process needs to support all three problem classes. The systems engineering process possesses the characteristics and strengths necessary to provide solutions for any problems in these classes that have realizable solutions. Since these problem classes have different initial conditions, the starting points are different.

However, the approach to developing a system to solve a problem is essentially the same across all three problem types. Systems engineering addresses all of them.

The Design Space: Three Systems

Figure 1

Every system design or improvement effort takes place in the context of three systems (Figure 1.) The most obvious is the system being designed. In the example problem, it is the system that will manage the images requested by the customers, taking them from inventory or procuring them from the collectors and sending them to the requesting customers.

The system being designed will function in the context of a larger system. In the example problem, the customers and the collectors reside outside the system to be designed. The customers interact with the system by requesting images and receiving the images requested. The system is created to "solve" their problem/need for images. The image collectors interact with the system by accepting tasking from the system and returning images in response. Both customers and collectors are part of the greater contextual system in which the system under design "lives and works." This contextual system is the second of the three systems.

The third and final system is the system that is used to design the system and bring it into being. This system is critical because it drives the quality and ultimate success of the design. This is the system that must understand the other two systems and must at the same time be "self-aware." It is only through this self-awareness that the design can take on its full measure of intentionality and manage the considerations manifested in the other two systems.

In one of the more interesting and helpful variants of the design space discussion, James Martin posits seven systems. He titles his article "The Seven Samurai of Systems Engineering: Dealing with the Complexity of 7 Interrelated Systems" as an allusion to the Japanese movie *The Seven Samurai*, in which seven samurai warriors fight to save a small Japanese village (http://www.incose.org/wma/library/docs/Seven_Samurai-Martin-paper-v040316a.pdf). Martin suggests that, properly employed, the seven systems he sees in the design space can become the key to design success.

Martin's systems are the Context System, the Intervention System, the Realization System, the Deployed System, the Collaborating System, the Sustainment System, and the Competing System. He summarizes their interactions as follows:

1. Context System contains a Problem.
2. Intervention System is intended to address Problem.
3. Realization System brings Intervening System into being.
4. Intervening System is a constituent of Realization System.

5. Realization System needs to understand Context System.
6. Realization System needs to understand the Modified Context System.
7. Realization System may need to develop or modify the Sustainment System.
8. Intervention System becomes Deployed System.
9. Context System becomes the Modified Context System.
10. Deployed System is contained in Context System. Deployed System collaborates with one or more Collaborating Systems.
11. Deployed System is sustained by Sustainment System.
12. Deployed System may cause new Problem.
13. Competing Systems may address the original Problem.
14. Competing Systems compete with Deployed System for resources and for the attention of users and operators.

Looking at the relationships and systems posited by Martin shows a clear mapping back to the simpler three-system model. The Context System in both Martin's model and the three-system model contains the problem from the outset and will contain the Deployed System once it is designed. Collaborating and Competing Systems also reside there. When the solution is deployed into the context, it will change the Context System. Understanding all of this is critical for the systems engineers to successfully create an acceptable design solution.

The whole issue of unintended consequences is an example of the failure to completely understand the Context System. Take as an example the introduction of nutria (a large South American rodent) as an "answer" to the need to produce quality fur more quickly and inexpensively. The following advertisement (Figure 2) frames the problem and solution that led to importing large numbers of nutria into the United States.

Figure 2

In the advertisement, nutria are touted as being productive (their fur next to mink in price); prolific (producing 15 to 20 young per year); easy to raise; inexpensive (costing only 1 ½ cents per day for food), as well as climate tolerant and disease resistant. However, it fails to address what turned out to be a major problem. The nutria reproduced rapidly and shortly "breached containment," escaping into the wild. Their reproduction rate, climate tolerance, and disease resistance made them formidable competitors in the wild. Soon they were driving out indigenous species and defoliating the habitats. Significant damage is being done to the areas where they have established themselves.

The "systems engineers" at companies like Cabana Nutria, Inc. understood the problem and addressed it directly. What they failed to understand and address was the Context System beyond the problem space. An effective solution to the problem created a much bigger problem as an unintended and unforeseen consequence of its application.

Even more insidious is the failure to account for the design system, what Martin calls the Realization System. Inattention to this system can cause a failure of discipline and rigor. It cannot be said too often in the world of systems engineering that we don't know what we don't know. A failure to use a disciplined or convergent process will lead to errors that will be completely transparent to those who make them. They are transparent precisely because of the failure to use the discipline that would allow the designer to catch them. Without a rigorous, disciplined system for design, there is no way to be sure that the design considers all aspects of the Context System.

The metaphor of three systems is simple, but its message is clear—systems design must consider not only the system being designed but its context and its method of design as well. Failure to take any of the three into account is a recipe for failure.

The Design Space: Boundaries

Once the systems engineer grasps the concept of the three systems, she must come to grips with the boundary between the system being designed and the system it will live in. The former is within the "control" of the design process, and the latter is simply present and must be adapted to.

The sample problem has an excellent illustration of the possibilities in the boundary question. The problem could present itself as an organization with an existing system of collectors desiring to develop a management system for its images that will increase the efficiency with which it can provide those images to customers (boundary 1 in Figure 3.) The same problem might arise when an organization has seen inefficiency between collector systems and their customers and recognized the business opportunity to bridge that gap by providing a way to reduce the inefficiency (boundary 2 below.) A third possibility might exist if the customer recognized that he could reduce his costs and wait time by managing the images already produced and seeking to engage the collectors for "new" images (boundary 3 below.)

In each case the system boundary is drawn differently for the design process.

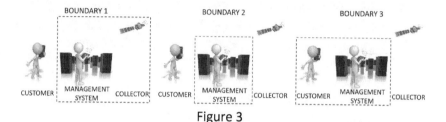

Figure 3

In each of these three cases, the degree of vertical integration helps determine the system boundaries. It is clear that it is very important to have a well-defined view of those boundaries, because no matter which boundary case applies to the situation confronting the design team, it is important to understand the nature of the problem in connecting the customers to the Geospatial Library and through it to the collectors. The design of those connections turns on the location of those boundaries.

The Process

The first task of the systems engineer is to develop a clear statement of the problem, setting out what issue or issues are being addressed by the proposed system. This involves working with others (especially system stakeholders and subject-matter experts) to identify the stated requirements that govern what would characterize an acceptable solution.

The systems engineer must provide design focus and facilitate proper and effective communication between the various subject-matter experts and the stakeholders. The systems engineer must have a broad knowledge base in order to understand the various disciplines involved in developing the system, to participate in and evaluate system-level design decisions, and to resolve system issues. Often some system requirements conflict with each other. When this happens, the systems engineer must resolve these conflicts in a way that does not lose sight of the system's purpose. The goal of the engineer is to develop a system that maximizes the strengths and benefits of the system while minimizing its flaws and weaknesses.

For illustrative purposes, we will present the systems engineering effort from a top-down perspective. Figure 4 illustrates the process of working across the domains.

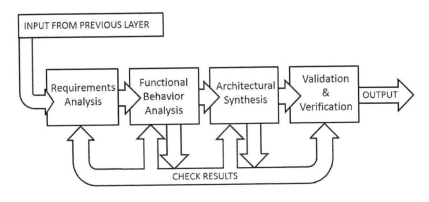

Figure 4

Once the problem is clearly defined, the process steps follow the flow in Figure 4. The originating requirements (which identify what the system will provide) are extracted from the source documentation, market studies, or other expressions of the system definition and analyzed. This analysis identifies numerous aspects of the desired system. Analyzing the requirements allows the systems engineer to define the system boundaries and identify what is inside and outside those boundaries. The definition of these boundaries—an often overlooked step—is critical to properly implementing the system. Any change in the system boundaries will affect the complexity and character of the way in which the system interacts and interfaces with the environment.

Specifying the functional requirements and the interactions of the system with the external entities is essential and leads directly to developing a clear picture of the system. It is critical that the functional requirements (the "what") are understood before attempting to define the implementation (the "how") of the system. Therefore, this analysis is repeated throughout the system design process to test the rigor and integrity of the system.

The "how" of the system is embodied in the functional behavior. This behavior is designed to meet the requirements as they have been laid out. Every requirement is the basis of one or more behaviors, and every behavior is based on one or more requirements. This is the backbone of the bidirectional traceability that will ultimately guarantee that the system design meets the system requirements.

In parallel with the functional behavior definition, what must be built to perform the needed behavior is derived through decomposing the system into components. The systems engineer then analyzes the constraints and allocates system behavior to the physical components. This leads to the specification of each system component. As a result of this allocation, the identification and definition of all interfaces between the physical parts of the system—including hardware, software, and people—can take place. These design decisions create the physical architecture of the system.

Once the physical architecture has been created and the behavior allocated to it, the system must be tested. Verification and validation are both aspects of that testing. The system is verified against the performance standards and specifications, and the design is validated against the requirements. In this way the engineering team can be certain that the design is indeed what is called for in satisfying the customers' needs.

It should be emphasized that these activities are performed concurrently or in sequence but not independently. The activities in one area influence, and are influenced, by the other activities.

This look at the systems engineering process uses a description that best fits top-down systems design. With suitable approach variations, the systems engineer can address reverse (or bottom-up) and middle-out systems engineering perspectives as well. For example, in conducting reverse engineering, the engineer would begin with the existing physical description and work from the physical representation and interfaces to ultimately derive the original system's requirements. Once these are obtained, the engineering process would proceed to incorporate the desired changes and enhancements as in a top-down design.

Domains

As described above, the work on the design proceeds in four domains: requirements, functional behavior, architecture, and verification and validation. Often these domains are treated as discrete efforts. In the classic approach, work in the domains proceeds in order, with the goal of finishing each one in turn. In this instance the process we have described moves sequentially through these domains.

Requirements

The problem-solving process generally begins with an exploration of the stakeholders' needs. This is a very high-level inquiry and results in a general statement of the system functionality. For example, in the sample problem it might be as simple as "We need some way to manage and deliver images from the Collectors to the Customers." Even at this very general point, it is possible to map the context system at a gross level.

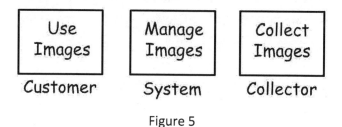

Figure 5

As the requirements process progresses, the requirements are made more specific. It is important to the systems engineering process that this increase in specificity goes hand-in-hand with the development of some increasingly granular concept of system behavior. That is because the requirements become the basis for behavior which is then allocated to architecture. Along the way there will be effects of behavior and/or architecture on each other and on the requirements. Attempts to drive the requirements completely to ground before proceeding to the behavior mapping or allocation to architecture make coping with those effects both costly and time consuming.

As the design is driven further and further into the details, the requirements develop specificity. As an example, consider the relatively high-level requirement that the system in the sample problem "provide continuous real-time support to the customers and the collection systems." The next level question becomes "What is continuous real-time support?" The requirement is refined to answer that question by defining it as "The system shall be unavailable no more than a total of 10 minutes per month." Each of the requirements is the basis of behavior that increases in specificity with the requirements.

Behavior

System behavior is concerned with two fundamental system characteristics: what the system must do in order to answer the customer's need and how well the system must perform these functions. In the example above what the system must "do" is be available to customers and collectors. The standard of performance (the "how well" it must perform) in this case is to the level of no more than 10 minutes of unavailability a month.

Describing behavior to meet requirements is constructing the system logic or the logical model. Often this is done in one step with the architecture design. The systems engineer attempts to allocate the requirements directly to the capability of an architectural component or components. Recognizing this as "doing" behavior in one step with architecture is probably being excessively charitable to that process. It is more appropriate to acknowledge that it simply skips the behavioral or logical step.

Separating the logical model from the physical model offers distinct advantages to the systems engineering team. The primary advantage is that the team does not have to assign behavior to components prematurely, something which complicates the system design effort. Separating these simplifies the design effort and enhances the likelihood of finding the best balance of functionality, performance, and component composition. Once the behavior is in place, it becomes appropriate to consider the architecture to which it is to be allocated.

Architecture

System architecture/synthesis is concerned with what physical structure offers the best balance—considering manufacturing, testing, support, and other factors—in answering the customer's need for the system. At its heart is the realizability of the system and its physical complexity. (Is it manufacturable, maintainable, and supportable?)

These aspects of the system strongly correlate with the system's behavior and the resulting partitioning and allocation of behavior to subordinate physical components. Architecture/synthesis tends to follow behavior in development (form follows function) rather than the other way around, because behavior more fully captures the features of what the system does. There is, however, an important "but."

Many systems have significant architectural constraints, which limit the systems engineer's choices regarding architectural composition. These constrain system behavior and therefore lead the way instead of following. They may even reach a level where it becomes clear that choices need to be made between competing requirements which impose mutually exclusive physical constraints. Such constraints are encountered at the point of architectural design and need to be translated back into the behavior (and sometimes even the requirements) domain (e.g., an aircraft whose specified operational range requires it to carry such a large load of fuel that it is too large for the short-field operations called for in the initial requirements).

Verification and Validation

At the conclusion of the systems development phase of the system life cycle, the customer must accept or reject the delivered system. Depending upon the situation, the customer may choose one or more methods to determine whether the system fulfills the requirements of the development contract. The goal in any case is to assure that the design process has converged on a complete and workable solution to the entire problem posed to the design team. The process for doing this is known as verification and validation. Verification and validation are the two aspects of answering the acceptance question. Both of these aspects need to be considered throughout the design and development processes. Program management and engineering management teams need to plan and address the measures that will lead to achieving customer acceptance.

Verification is a "quality" process used to evaluate whether or not a product, service, or system complies with particular regulations, specifications, or conditions. Verification may occur anywhere in the system's life cycle. Verification is often an internal process, but external and independent verifications can also occur.

While validation is the process of establishing necessary and sufficient evidence that a product, service, or system satisfies its established requirements, formal validation often includes the confirmation of fitness for use from the viewpoints of customers, end users, and other product stakeholders as an acceptance criterion. The ultimate validation question becomes: "Does this system, as built, satisfy the needs which drove the instigation of the design project?"

NOTE: More and more systems engineering teams think of the process of measuring the efficacy of the system design as Test and Evaluation (T&E). This concept arises from the software development world and seeks to document the scope, content, and methodology for test activities. It is embodied in a test plan which describes the test activities of the subsystem integration test, the system test, the user acceptance test, and the security test in progressively higher levels of detail as the system is developed. (As we shall see, this makes T&E a natural fit with the layered MBSE approach.) Although T&E is different from classic V&V, they are similar enough that for the purposes of this primer we will treat them as the same.

Communication

It is easy to see that these four domains cover a variety of disciplines. From gathering the requirements from an often diverse community of system users and owners to specifying technical architectures that will cover complex capabilities, the systems engineering team must be able to communicate the problem and the potential solutions in ways that will be universally understood.

In the sample problem, the customers who use the images will be experts at interpreting the information in the images, visualizing terrain and recognizing infrastructure and human activity. They will understand what images they need and what those images should contain. They will not necessarily be familiar with the technology necessary to allow them to phone, fax, deliver, or directly request the images they need. The designers of the library will understand how to use the image data and metadata to organize, store, and retrieve the images but will not know the technical details of the alternative ways in which the images can be brought into the library from the collectors' information stream. Bridging these gaps and many others is the job of the systems engineering team.

This requires careful and nuanced communication. There must be attention to language, and a common understanding of the use of language must be developed across the design and the disciplines involved in creating it. This coordination and communication challenge makes the effective practice of systems engineering a challenge in both the management and technical arenas.

The successful systems engineering team must maintain a systems view while moving through the four domains. Stakeholder needs are the source of system requirements. Those requirements become the basis of system behavior. That behavior is allocated to the physical architecture, which is then judged back against the requirements. Along the way, the team must craft that solution which best fits the context for it and do so using a disciplined and effective design process.

WHAT IS A MODEL?

Models are common to human experience as aids for understanding the way the world works. Everyone has experience with some form of model and therefore has some preconceived notions of what constitutes a "good" model. Children's toys are simple models of the world around them. Toy cars, trains, and dolls all typically characterize forms, playing on the child's ability to link imagination (an abstract representation) to a real object. In this sense the word *model* means a physical representation of an abstract idea.

Models span a spectrum running from form to function. On one end are tangible, visible models like a child's plastic toy airplane. It mimics, or models, the physical appearance of the object (the full-sized airplane) that it represents. It doesn't fly—or if it does it doesn't do so in the way the actual object does. It models the form of the plane but not the function.

On the other end of the spectrum are model forms existing only as sets of equations or simulations implementing the equations. Rather than visually representing the reality behind them, these models allow us to examine such things as the behavior of the object being modeled. One common characteristic of such models is that they capture or emphasize only certain properties of interest in the modeled object, while the fidelity of the model to the actual object is intentionally reduced or limited in other ways.

In the world of engineering design, models connect the idea behind a design solution with its implementation as a real system. These models attempt to represent the entities of the engineering problem (opportunities) and their relationships to each other and connect them to the proposed solution or existing mechanism that addresses the problem. The model used in this way is the centerpiece of MBSE.

Four Elements of a Model

There are four elements of such a model: language, structure, argumentation, and presentation.

Language—The model is seen in terms of language. The system definition language (SDL) expresses and represents the model clearly, so that understanding and insight can arise. This is critical to successful system design. The system definition language must be clear and unambiguous in order to depict the model accurately and understandably.

Structure—The model must have structure. This allows the model to capture system behavior by clearly describing the relationships of the system's entities to each other.

Argumentation—The purpose of the model is to represent the system design in such a way that the design team can demonstrate that the system accomplishes the purposes for which it is designed. Therefore the model must be capable of making the critical "argument" that the system fulfills the stakeholders' requirements.

Presentation—Not only must the system be capable of making that argument, but it must include some mechanism of showing or "presenting" the argument in a way that can be seen and understood.

These elements, language, structure, argumentation, and presentation, give the MBSE model what it needs to serve the purpose of testing the system design solution against the requirements in a way that proves its fitness and presents that proof for all to see. This is the distinguishing value of the model.

Characteristics of a Model

There are four characteristics common to successful system models. These are order, the power to demonstrate and persuade, integrity and consistency, and the ability to provide insight into both the problem and its potential solutions.

Order—Order allows the design team to attack the problem in a coherent and consistent manner leading to a viable solution. The model provides the order that becomes the framework for this effort.

Power to Demonstrate and Persuade—By representing the relevant behaviors in proper relationship to the system entities, the model allows the designer to see and demonstrate the necessary system behavior. This becomes persuasive in making the case that a given solution answers the needs that drive the design of the system.

Integrity and Consistency—Ambiguity and inconsistency in the system design lead to design flaws which, in turn, harm the credibility of the argument that the system design meets the needs it was designed to meet. The model must, therefore, provide the integrity and consistency that lead to a sound solution.

Insight—The model provides insight into the system problem facing the design team as well as the potential design solutions. By the model's representation of system behaviors and relationships, the design team is able to gain insight into the comparative advantages of different approaches to solving the design problem at hand.

Caveat: A Set of Views Is NOT a Model

Various graphical and textual views derived from the true systems model are sometimes treated as if they were themselves models. However, these are, at most, viewable projections of the underlying model. That is, they contain some subset of entities, attributes, and relationships presented so that the engineer, reader, or reviewer gains insight into a particular aspect or aspects of the system design. Graphical or textual views, in themselves, are not sufficient to constitute a model. They are, rather, expressions of the model being represented.

To be a true model, the system model needs to manage the depth, breadth, and associated boundary conditions of the system. This is not possible with a view or even a set of views. Views are a valuable tool for understanding, analyzing, and communicating the model. Some sets of views even offer a broad understanding of many system aspects. But the views themselves are not a model.

Language: The Systems Model Is Language-Based

The relationship between the language expressing the model and the meaning conveyed in the model is critically important. Language is critical to disciplined systems design. The ambiguity and lack of clarity that are so often present in design efforts can have crippling results which can render a system design useless. The need for a clear, unambiguous systems definition language is only enlarged by the presence of a diversity of disciplinary experts required to assist in a complete design. The language will include both the symbolic representation of system concepts and the graphic views and representations that are used to convey the functions and behaviors embodied in the system. After choosing a particular reference entity of a given class, the use of the definition language enables the engineering team to ask the right question at the right time. A tremendous advantage for resolving issues, this avoids unnecessary or inappropriate work. A common language that can give full expression to the system in its entirety is essential to a successful design and to making the case to the stakeholders that the design actually meets the requirements posed in the problem that drives the design.

The model must be much more than one or more graphical representations. It must take on the difficult tasks of representing the system's relationships in a way that assures traceability and the consistency of boundary conditions across the domains. The model is therefore captured in a language in a way that allows the engineer to determine and communicate the system characteristics. These characteristics drive the way the system's components interact within the system and with the system's external environment.

Another aspect of a system as a whole is that it cannot be divided into independent parts without losing some of its essential characteristics. Thus, a system's essential defining properties are the products of the interactions of its parts, not the sum of the actions of the parts considered separately. This means that a successful system language must be able to capture these essential interactions in a way that accurately depicts this synergy.

Sometimes there are kinds of behavior and properties that the system must exclude. These exclusions are as much a part of the system definition as those that are included. Safety and security properties fall into this realm. The system must not be unsafe to users. It must not be vulnerable to specific threats. The model must be clear in expressing whether or not these system properties will or will not be present. This is particularly true because the properties of individual components are not necessarily present in the system. Graphics by themselves have only a limited ability to convey these characteristics and assurances. This is one of the fundamental reasons for needing an expression of the model that extends beyond mere graphic representations.

A model is an integrated expression of the system using the system definition language (SDL). It comprises source or originating properties (e.g., context, purpose, environment, and other constraints), physical properties (e.g., size, weight, power), behavioral properties (e.g., events, time sequencing of observables, execution conditions, performance), relevant analytical and test information, and the relationships between these system entities.

Other characteristics are necessary for a successful systems language such as the system definition language. It must be relatively easy for a diverse population to understand it, while at the same time it must be able to deal with the necessary levels of abstraction. Not everyone possesses the knowledge to understand every nuance of the system model. Therefore, the SDL language needs to use a basic vocabulary without a multiplicity of meanings.

With SDL, the specialty language of domain experts is avoided; yet domain experts can easily relate the SDL to their domain of expertise. The mapping in Figure 7 (page 38) represents the relationship of the parts of speech from common language to model-based systems engineering SDL.

A Primer for Model-Based Systems Engineering

English Equivalent	System Definition Language	MBSE Example
Noun	Entity	**Requirement:** Place Orders **Function:** Collect Images
Verb	Relationship	Requirement is the **basis of** Function
Adjective	Attribute	**Description**
Adverb	Relationship Attribute	Function consumes Resource **Amount** of Resource being consumed
N/A	Structure	Activity Diagram Enhanced Functional Flow Block Diagram

Figure 6

As stated earlier, the model concept and the SDL are interdependent. The language must possess a minimum set of nouns sufficient to identify the objects typically encountered in performing systems engineering. It must include such noun classes as Component, Function, and Requirement. Each noun class possesses a set of adjectives (attributes) that refines and adds depth to each particular class instance. An examination of Figure 7 shows the noun classes associated within the operational and system portions of a DoDAF architecture.

In the case of DoDAF, the Architecture class acts as a key element. It brings the physical natures of the operational and system sides together. Thus, in a physical sense, it is clear that a particular Architecture entity provides the context for understanding how a set of operational entities and a corresponding set of system entities relate.

Individual relations constitute the verb forms of the SDL. Thus, once the set of noun classes is constituted, the relationships among them lead to the identification of the set of relationship pairs (verbs) needed in the SDL.

A Primer for Model-Based Systems Engineering

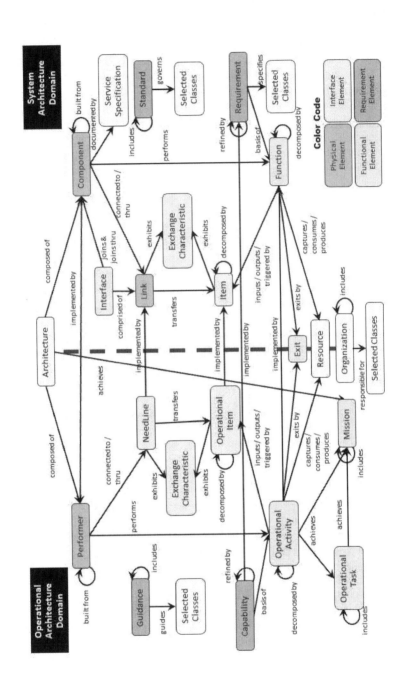

Figure 7

The directional lines in Figure 7 illustrate some of the primary SDL relations. These relations are directional. This means that in forming a "sentence," the noun class at the arrow's tail is the subject of the sentence and the noun class at the arrow's head is the direct object (e.g., Architecture is "composed of" a Component). Such a sentence expresses a "relationship."

The types of problems encountered determine the types of nouns, verbs, adjectives, and adverbs needed in the SDL. The solvable problems need language entities to define both the problem and solution. The fundamental structure of the SDL does not change, but vocabulary (nouns, verbs, adjectives, and adverbs) should be tailored to the domain. The definition of the system definition language structure and vocabulary is critical to having a successful system design.

Language of Behavior

The need for a language to express the system design applies to all aspects of the design. This includes the representation of behavior. The language of behavior is graphical. It must reflect the time-dependent and time-independent aspects of system behavior, the sequencing of functions, resource management, interfaces, and control behavior.

The criteria necessary for showing behavior graphically require that the graphical language (notation) possesses certain characteristics. These characteristics are:

1. The ability to capture process flow and control
2. The ability to capture observables
3. Understandability
4. Executability
5. The ability to preserve behavior across:
    a. Decomposition
    b. Aggregation
    c. Allocation

The system definition language functions and items are both decomposable. Items represent the inputs and outputs of functions—the observables. Any system may be described in terms of "black boxes." A black-box view just addresses the inputs and outputs to the system, including their sequencing and timing. Such a system is also decomposable into a set of black boxes.

It is important that any decomposition (or aggregation) not lose the behavioral effects of the layer from which it is based. For example, the behavioral consequences of a trigger (an input that enables the function to begin) should be preserved into the next layer upon decomposition and not be "lost." Preservation of behavior is, therefore, an important property of both the graphical language and the system definition language.

For example, in the Geospatial Library design, one of the high-level system requirements calls for the system to "accept requests from certified customers." As the design is fleshed out, it becomes apparent that customers need to make requests in a variety of media formats. The requirement to accept requests from certified customers must encompass these formats, so it becomes a requirement to "accept requests from certified customers via any of the following media: (1) hardcopy forms, (2) verbal, (3) phone-verbal, (4) phone–electronic file, and (5) PC diskette–electronic file."

Although the behavior "accept request" has been decomposed into the several differing behaviors necessary to accept the various request formats, this decomposition cannot change the trigger-response behavioral consequences of accepting the customer request.

Behavior preservation under both decomposition and aggregation allows the logical design to be consistent through each design layer. If the logical model cannot remain consistent within each layer as well as across the layers, the system definition language is critically flawed. Preservation of behavior under allocation is necessary for the logical model both as a whole and across the layers. A simulation of the integrated logical model should give identical results to a simulation of the allocated model.

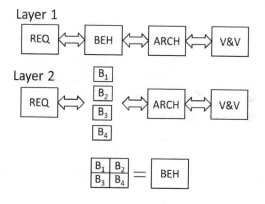

Figure 8

Because the process must be traceable, all observables must be preserved whether under decomposition or aggregation. Inputs and outputs, input and output sequencing, and the number of and conditions for exits and performance all need preservation. For example, as is illustrated in Figure 8 above, when Behavior (BEH) is decomposed from one layer to the next (BEH becomes B1–B4) the functional equivalence (B1–B4 = BEH) must be preserved. Performance characteristics under decomposition need to be carefully managed. Performance measures may change at different levels of decomposition and performance characteristics, especially time performance values, may need decomposition as well.

Managing Complexity with Language

Separating the functional/behavioral domain from the architecture/synthesis domain is one of several means of managing complexity in model-based systems engineering. The logical model is one that, over time, changes little, if at all; while the physical model changes rapidly due to technological and other advances. Using the principle of allocation, many alternative physical configurations may be tried to find the best balance among a diverse set of criteria. Later in the system life cycle, if the physical structure needs to change, it becomes relatively easy using the model to discover what functions are affected by a physical change and vice versa. Conversely, if a requirement change occurs, then its impact on the functional model and physical models is also easily discovered because of the traceability afforded by the system definition language.

There are a number of diagrams currently used to support the behavioral domain in systems engineering. The more prevalent views are shown in Figure 9. Several structure diagrams capture the logical sequencing of functions, events, and control flow. These are the Functional Flow Block Diagram (FFBD), the Enhanced Functional Flow Block Diagram (EFFBD), and the Activity Diagram.

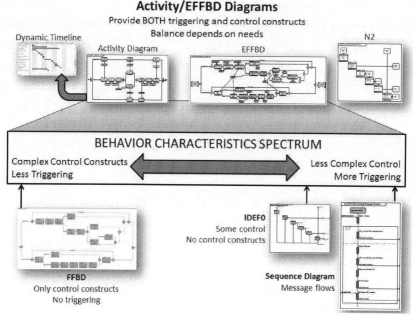

Figure 9

Functional flow block diagrams represent one limit on the behavior spectrum. They show control but no data or data flows. At the other end of the behavior spectrum, data flow diagrams, N2 charts, and sequence diagrams show only data or data flow but no control. However, the behavior diagrams reveal both control and data. These diagram formats span the full spectrum of behavior.

Though limited in content, data-oriented views do serve a valuable purpose. Most uses of diagrams on the data side of the behavior spectrum have found their use in instances where an entity, particularly a human, provides the control and the actions are event-driven, with limited need to communicate across events. In such cases, the results may be captured in databases, and dashboards are used for presentation.

State diagrams are yet another type of graphical representation that may be encountered. States represent a quiescent node in the sequence of system events. A control-like triggering event causes a state change, where the transition from one state to another captures the transformational change of something in the system and its rest at the new state. These state flows are equivalent to the control flows within a behavior diagram with the quiescent points in the system corresponding to where a branch function awaits a trigger.

With a rich set of possible representations available, the key is to select the diagram type best suited for the analytical or communication need. However, it is essential that the views be derived from an authoritative underlying model lest they—and the system—become inconsistent.

Graphical views can be generated from the model using the relationships and attributes to define the diagram's structure. This allows members of the engineering team and others to grasp the logical system design. The graphical language used in model-based systems engineering permits simulation through a discrete event simulator. The simulator, integrated with the system repository, minimizes logical inconsistencies in the model. This improves the consistency and quality of any specifications, design documents, or other artifacts generated from the system model.

Structure: The Model Expresses System Relationships

The meaningful set of relationships in the model must be expressible in both global and local contexts. The global context contains the system and its externals. The local context includes a subordinate system component and its more localized externals. These localized externals are typically other system components (system components are external to each other but internal to the system). However, particular instances may include one or more of the system's externals as an external to a component.

Control Constructs

Behavior structure diagrams provide the means to develop the logic of what a system or other entity does. This logical representation is not unique, but it serves to generate the inputs and outputs that an observer external to the system sees over some observation period. The time relationships and sequencing of these inputs and outputs are part of the system's requirement structure.

A structure diagram contains at least one branch. A branch is the diagrammatic language entity that organizes functions and other control constructs into a logical order. This provides the sequencing for functional execution. The control constructs for control of that sequencing, execution path selection, and decision logic are determined by the placement of the functions on this branch. Sub-branching occurs through the several types of control structures.

The branching is the basic representation of the behavioral logic. It occurs due to the action of a variety of control constructs. The following list summarizes each construct and its effect on behavior.

- **Sequence**—The left-to-right ordering of the functions on the enablement branch defines the sequence of execution, unless modified by a control structure. In Figure 10 below, information requests must be made before products can be accepted.

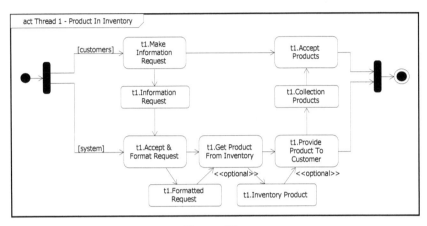

Figure 10

- **Concurrency (or AND) construct**—When the process reaches the first AND node, it will split and enable the first functions on each of the parallel branches in the AND construct. The process will not proceed past the second or terminating AND node of the concurrency construct until all functions on all branches have completed execution.

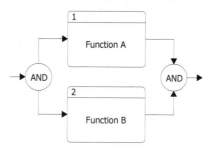

Figure 11

- **Select (or OR) construct**—One branch is chosen out of the possible enablement branches and all other possible branches (and the functions and constructs on them) will be excluded (an exclusive OR). The selection of the branch may occur by some event occurring, scheduling scheme, or business rule satisfied. Once all of the behavior logic and functions have completed execution on the chosen branch, control passes on to the next function or construct after the second or terminating OR node.

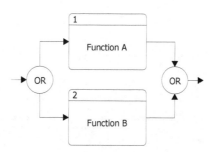

Figure 12

- **Multiple Exit Function**—Logic or business rules inside the function possessing multiple exits will choose one of the possible exit paths, each representing a unique enablement branch. The choice of one branch of a multiple exit function behaves similarly to a Select (or OR) construct, hence the terminating OR node.

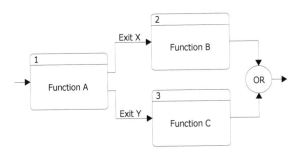

Figure 13

- **Iterate (IT)**—The collection of functions and constructs between the IT nodes of the construct will repeat for a specified number of times, repeat across a specified set of objects, or repeat at a specified frequency.

Figure 14

A Primer for Model-Based Systems Engineering

- **Loop (LP)**—The collection of functions and constructs between the LP nodes will repeat until some exit condition occurs.

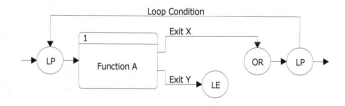

Figure 15

- **Loop Exit**—A loop exit specifies that control (vis-à-vis the imaginary enablement token) is to be passed on to the branch immediately after the second or terminating LP node. It occurs on one of the branches of an OR construct within an LP node.

- **Replicate**—Specifies a set of behaviors between the RP nodes that is instantiated in multiple and independent cases. This is analogous to placing multiple identical branches on an AND construct.

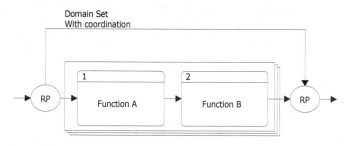

Figure 16

- **Exit**—Specifies the identity of a node corresponding to a parent-level exit condition for the entity on which the diagram has been opened. Thus, the Exit node identifies the desired path on the parent graph to be taken as a result of reaching the exit node on the child behavior diagram. The Exit node, therefore, represents the mapping between internally represented exit behavior and the externally represented exit branches for the parent function.

Behavior is represented in activity or functional flow diagrams. Both functional flow block diagrams—FFBDs and EFFBDs—and activity diagrams are read from left to right. One helpful aid is to follow an imaginary "enablement token" that starts on the far left of the diagram. At an initial AND branch symbol the imaginary token splits to follow each parallel branch. It will move from left to right along the horizontal "enabled branches" and enables, or allows, a function to execute when the token reaches it. At the terminating AND branch symbol, all the subordinate imaginary tokens are rejoined before the token continues its journey.

At an initial OR branch symbol, the imaginary token advances along only one branch (an exclusive OR). The selection of which branch to follow is based on some event occurring (perhaps as scheduled) or upon a business rule being satisfied. Once all the branch functions have executed, the imaginary token continues its journey.

In Figure 17, the product requested is NOT in inventory, so the token takes the lower branch (Not in Inventory). It proceeds through Prioritize Request and Determine Collector Mix to the AND branching. There it divides to both Notify User of Estimated Schedule and Task Collectors in parallel. The token then moves to Accept and Format Collector Products and Put Product in Inventory. It then proceeds to Get Product from Inventory and divides again at the AND branching to both Provide Product to Customer and Evaluate Products vs. Request. The Product is OK, so the process continues on the lower of the OR branches and reunites to finish the process.

A Primer for Model-Based Systems Engineering

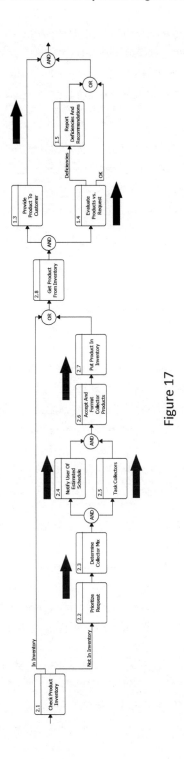

Figure 17

Branches are the way functional behavior is specified using an EFFBD or activity diagram. Each of the different control constructs has a unique effect on the flow of the enablement token, which is how control constructs specify the control logic.

These constructs serve as building blocks for the basic system logic. However, there are two other modeling concepts that affect functional behavior. These two modeling concepts are the triggering of functions and provisioning of resources.

Triggers and resources are terms used in the system definition language. Triggering is the determination of when a function may execute on a branch once that branch is enabled. Enablement in this sense means that all prior functions have completed execution. A triggered function must wait for one or more incoming, or triggering, items to be received before it may execute. This is important when functions are allocated to different components. Triggers serve as a means of execution control under allocation. It is the means by which a functionally allocated component knows when to begin operation. (Triggering is often highlighted in sequence diagrams, showing interactions between components but without the clear specification of control.)

Resources also affect function execution. For example, if a function exists to fire a missile, the firing (function) is ineffective if there is no missile (resources) to fire. Modeling resources accounts for a function becoming resource-starved (no missile to fire) or having diminished performance where its necessary resources are absent or degraded.

The activity diagram and enhanced functional flow block diagram span the behavior spectrum and communicate the logical steps that yield a given class of output for a given class of input. The ordering of functions, which either perform or contribute to the transformation, is important; just as word ordering in an English sentence is important. Changing the order of the functions, in general, changes the outcome; just as changing the triggering events changes the ordering of the outcomes. The result is that the graphical language is a special case of the more general system definition language.

Argument: The Model Is Used to "Prove" the Concept of the Design

The engineering team may consider different allocation schemes to address any given engineering issue. Each alternative offering will have certain benefits. The systems engineering team examines, analyzes, and tests these different allocations, as necessary, to achieve the best fit of the benefits to the system. The allocation is measured with consideration to a number of factors, including the system requirements, the number and complexity of interfaces, and life cycle costs, just to name a few. This analysis allows the team to verify and select the alternative that offers the best balance of performance, functionality, and usability in satisfying the customer's need. From that analysis, the engineering team can create the picture needed as the basis for verification and validation.

Bidirectional relationship sentences are used to reveal the system's interdependencies. (For example, A decomposes B, implying that B is decomposed by A.) Such sentences advance the model's description and definition. Model advancement means that the system unfolds in detail as the story-line progresses.

In contrast to the world of document-based engineering, the model's reverse path allows the systems engineering team and other stakeholders to unravel how the team reached a particular point in the design effort. Horizontal linkages reveal interrelated entities at the same complexity level. Vertical linkages reveal relationships among various layers of abstraction. The ability to express these system relationships meaningfully allows the systems engineering team to manage the complexity of the design and to identify and evaluate impacts on other system entities. Document-based approaches do not allow this (e.g., we cannot easily read a book in reverse). This gives the model-based approach a significant advantage over document-based approaches.

This advantage results because, with models, the system design is advanced through levels of increasing detail and demonstrates within itself how and why the system design will satisfy the stakeholders' needs. Therefore, a model is much more than a graphic, a series of graphics, a set of tables, simulation results, or even a collection of such things.

Presentation: The Model Must Be "Viewable"

This is the aspect of the model that places particular value on graphical views. It is one thing to describe the model abstractly and quite another to make the description real. For most of us, this means some way to visualize the model in three-dimensional space. Graphical views are helpful in prompting that visualization.

All views are filtered for relevance to their purpose. No one view contains the entire set of information. In fact, an attempt to do that would rob the view of its power to make a particular "statement" about the model by cluttering it up with information not relevant to its purpose.

Graphical Representations

The combination of the system design and the graphical languages allows the model repository to generate a variety of graphical views. Each view enhances and/or emphasizes some system characteristics and suppresses others, giving the engineering team a means of gaining different insights into the system and its design. The functional flow block diagram (FFBD) represents one limit of the behavior spectrum—where only the control or logical structure of some portion of the functional model is presented.

Figure 18 on page 56 shows a functional flow block diagram. In this instance, there are two primary branches arranged as concurrencies (a 2-branch AND construct). Each branch acts independently of the other and may be activated by triggers, but triggers are unseen in this view. Both branches need to complete before exiting the behavior logic (the terminating AND on the far right of the diagram). Rectangular icons represent functions on this diagram. Also in this diagram are two functions employing multiple exits. This is because some decision occurs within these two functions.

A Primer for Model-Based Systems Engineering

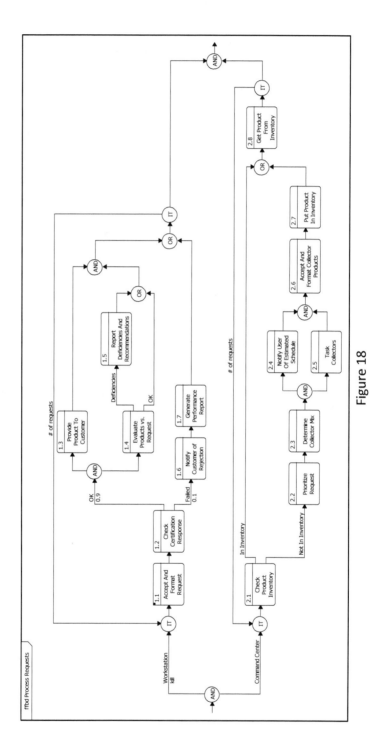

Figure 18

The view depicted in Figure 19 is a sequence diagram. For simplicity's sake, it shows thread 1 – the behavior that occurs if we assume the product is available in the inventory. (The corresponding activity diagram is shown in Figure 10 on page 46.) The sequence diagram focuses on interactions, clearly showing the triggering between functions (e.g., "Information Request" and "Collection Products") allocated to different components. In this view, time progresses down instead of across the diagram.

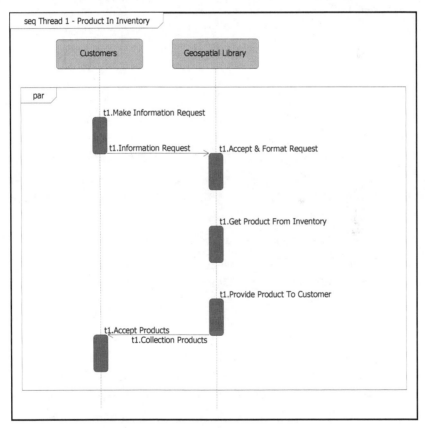

Figure 19

Figure 20 shows an activity diagram. In this instance, we see the same logic flow as in the functional flow block diagram (Figure 18), but with Items (inputs, outputs, and triggers) added. For example, note that the trigger "information request" appears above the function "Accept Request" in the upper left of the workstation branch in the activity diagram but *not* in the FFBD. This information is shown on the enhanced functional flow block diagram (EFFBD) shown in Figure 21 (page 60) as well.

In the activity diagram the trigger items point to the corresponding functions without an <<optional>> tag. In the EFFBD the trigger items and their inputs to functions possess a double-headed arrow. These icon characteristics indicate that the item is a trigger and influences the execution of the logical model. The first function on the top branch of the first AND construct (Accept and Format Request) has a triggering item (information request), but this item's icon does not have any line indicating its origin. Graphically, this means that this particular item was output from a function beyond the current graphic boundary. In this case it came from the Customer, who is not shown within the boundary of the diagram.

Upon execution, this Function outputs an item (formatted request.) This in turn becomes the trigger for the Check Product Inventory function on the lower AND branch. This shows that the execution of these two branches is not independent, as might appear from observing the functional flow block diagram alone. Clearly, the execution of the logical model depends highly upon triggering items.

A Primer for Model-Based Systems Engineering

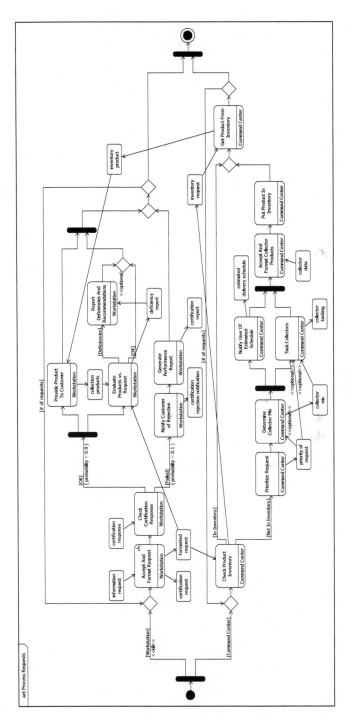

Figure 20

A Primer for Model-Based Systems Engineering

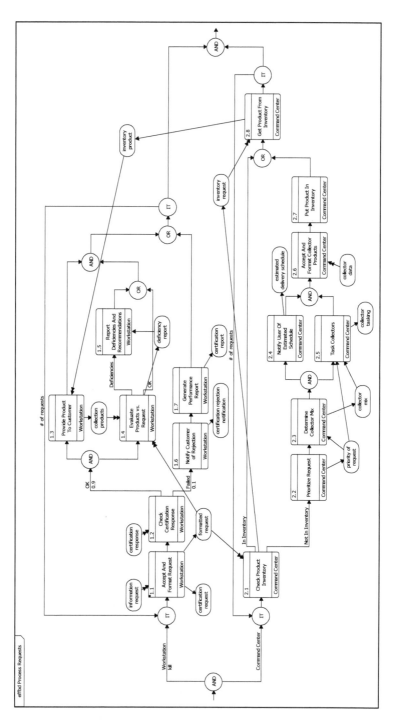

Figure 21

Another important property of both the enhanced and functional flow block diagrams is that time is a component of the view. Each function has a duration and the input and output items are the primary observables. Therefore, their sequencing in time may be seen using the model repository's discrete event simulator.

Because the functions in an enhanced or functional flow block diagram are the decomposition of a higher-level function, they may also be represented by a functional hierarchy diagram. Figure 22 (page 62) shows the functional hierarchy for the behavior diagrams shown in the diagrams above. The hierarchy diagram shows the depth of the functional structure, starting from a source function for the hierarchy. In addition, by selecting from a set of valid relationships for the class of the starting entity, a hierarchy diagram reveals the relationship paths and depth of the model for the selected relationship set. This diagram is helpful for revealing unfinished portions of the model as well as any miswiring of the model.

A Primer for Model-Based Systems Engineering

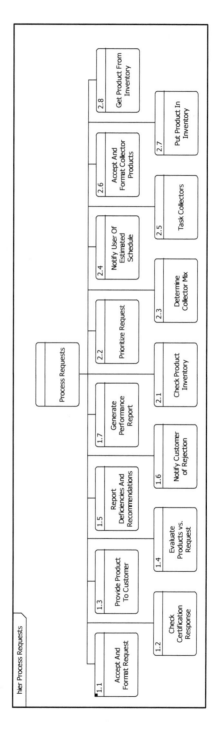

Figure 22

A Primer for Model-Based Systems Engineering

Another diagram helpful for analysis in both the functional/behavior and architecture/synthesis domains is the N2 (*N*-squared) Diagram. The N2 diagram simply places the functions of a behavior diagram along the principal diagonal and the respective inputs/triggers and outputs in the off-diagonal cells. Figure 23 (page 64) shows an N2 diagram. Horizontally oriented line segments between functions and items represent outputs, and vertically directed line segments between functions and items represent inputs or triggers. Any items appearing in the top row of the diagram represent items arising from an external function or functions. Correspondingly, any items appearing in the far-right column represent outputs going to external functions. Two aspects of this diagram are of interest to the engineering team. The first is that time is not considered, and the second is that functional interfaces are discerned. This diagram emphasizes data and provides insight into allocation and interfaces.

The use of a model provides a disciplined framework in which to construct the system. This is the case in the architectural domain as well. Just as behavior is described and represented graphically, so is physical architecture.

By looking at the model, the engineering team can "see" behaviors and architectures and test them against the system requirements. Although the various representations are powerful tools for understanding and communicating the model, they are not the model itself. They act as filters and containers for the information that the model embodies. But it is the language, structure, presentation, and argument that combine to make the model the tool of choice in the field of system design.

A Primer for Model-Based Systems Engineering

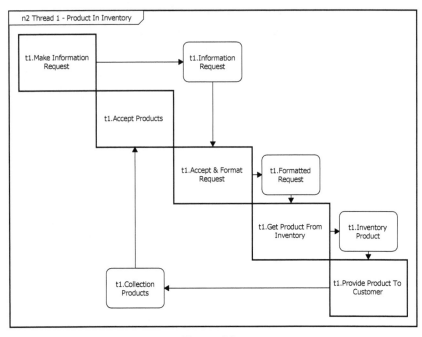

Figure 23

WHAT IS MODEL-BASED SYSTEMS ENGINEERING?

Model-based systems engineering (MBSE) is fundamentally a thought process. It provides the framework to allow the systems engineering team to be effective and consistent right from the start of any project. At the same time, it is flexible enough to allow the "thought" process to adapt to special constraints or circumstances present in the problem.

There are major advantages arising from using models as the basis of systems engineering. Models thoroughly consider the entire engineering problem, use a consistent language to describe the problem and the solution, produce a coherently designed solution, and comprehensively and verifiably answer all the system requirements posed by the problem. These traits of model-based systems engineering are significant advantages when seeking a solution to the systems design problem at hand.

The MBSE approach that is the subject of this primer is based on a "layered" process of analyzing and solving systems design problems. Developed by Jim Long, Marge Dyer, and Mack Alford, along with a succession of colleagues and associates, it has developed in a way that has now been named STRATA™ by Vitech Corporation. The name calls out the central principle of the method—Strategic Layers—because the method is built around handling the problem in layers of increasing granularity in order to converge strategically on the solution.

Beginning at the highest, most general level, the problem statement (system-level requirements set) is analyzed and translated into functional behaviors that the system must perform to fulfill the requirements. These behaviors are allocated to physical components (collectively an architecture) that provide the means for performance. The architecture is then tested to see that its performance answers the requirements.

As the system is developed in increasing detail, or "granularity," a layered structure takes shape. The engineering process follows these layers, drilling deeper and deeper into the system design. Every iteration of the systems engineering process increases the level of specificity, removes ambiguity, and resolves unknowns. The domains (Requirements, Functional Behavior, Architecture, and Validation and Verification) are all addressed in context at the level of increased detail as each successive layer is peeled back.

Instead of driving the work in each domain to completion (as in the traditional approach), layered MBSE works through each domain in each layer. The two methods are compared below in a simplified visualization.

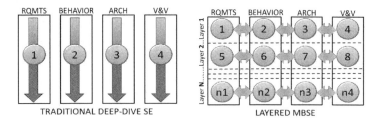

Figure 24

The visualization is simplified in both cases because it does not show any iteration between domains in the traditional approach nor does it show the iteration between layers in the MBSE methodology. In the traditional systems engineering approach, the iteration among the domains usually takes the form of rework and revisiting domains. This is expensive and always unplanned. The idea is to finish each domain and move on to the next without needing to return.

In MBSE the iteration can be confined to an adjacent layer and is easily achieved without interfering with the intended design progress. This represents a huge advantage over the traditional approach. Revisiting the domains often means restarting the design process (or a significant portion of it) in order to account for the changes made in that domain during the revisit. This causes reopening the design and reworking areas that have already been covered. None of that is necessary with the planned and incremental iterations that are part of the workflow in the layered MBSE method.

The layer-by-layer approach of MBSE assures that the domains are considered in context. One of the critical system design mistakes is losing the system context. This happens as a natural outgrowth of thinking about the design analytically—that is, by tearing it apart into its components and focusing at that level. The work of the late Russell Ackoff, a University of Pennsylvania business professor and pioneering "systems thinker," reminds us that we must also think "synthetically," holding the entire system in view and considering it as a whole. Because a system is, at its root, more than the simple sum of its parts, we cannot afford to lose the systems view or we lose the essence of the system itself. (See, e.g., R. Ackoff with H. Addison and A. Carey, *Systems Thinking for Curious Managers*, Triarchy Press, 2010.)

Approaches that involve "deep dives" into one area (e.g., requirements) run the substantial risk of obscuring the systemic risks incurred when the complex relationships between domains are not fully considered. The central power of the MBSE approach lies in its careful and complete consideration of the system design in an orderly and systematic fashion. This can happen only through the orderly process and excellent communication which must be the hallmarks of an effective systems engineering process.

This approach to solving the problem in layers is the heart of MBSE. With this approach (addressing all domains in each layer), there comes an assurance that all aspects of the engineering problem at hand are addressed completely and consistently. The layers and their interrelationships also provide a solution that can be easily verified and validated against the needs that created the problem. By stripping away the layers as we might "peel an onion," we can be assured that we have indeed addressed the problem in a meaningful and productive way.

As the layers are peeled away, the process converges on the solution. The discipline and rigor of the MBSE process assure that, like the peeled onion, the design never finds that the next layer is larger than the last. The process instead converges on the solution naturally as a product of the MBSE process.

Requirements for a Systems Engineering Process

In order for such a solution-seeking process to be effective, it must satisfy some fundamental requirements. MBSE answers these requirements quite well.

Requirement 1: The process must consistently lead to the development of successful systems.

MBSE is a coherent and comprehensive means of consistently arriving at a realizable system effectively and efficiently. By engineering the system horizontally in layers and completing all the systems engineering activities at one layer before decomposing/elaborating in the next layer, the MBSE engineering team advances the design from layer to layer. In this way, MBSE converges layer by layer on a system solution that successfully meets the needs behind the development process.

As the work progresses through the layers, the engineering activities performed in each domain change emphases. In layer 1, work focuses on the requirements domain along with verification and validation, if done properly. Work in layer 2 emphasizes the functional/behavior domain. Succeeding layers are a balance between the functional/behavior and architecture/synthesis domains. Refinements in system control, error handling, and resource management are added at each succeeding layer.

Preliminary specifications are available at the conclusion of each layer's activities. Therefore, the MBSE modeling process yields meaningful draft specifications throughout development, a distinct advantage over conventional approaches. Doing systems engineering in layers makes the MBSE process virtually fail-safe. Should external constraints force the process to stop at any given layer, the draft solution is complete to the level of the last layer finished.

A convergent process like STRATA uses this disciplined, ordered process to make the best choices under the given conditions. This is not to say that any particular decision is not flawed. But where flaws exist, any correction of them can occur without dire programmatic consequences. In addition, by framing and developing the solutions in layers, STRATA ensures that assumptions, boundaries, interfaces, functions, and architectures are convergent, consistent, and complete.

The convergent process is the means of avoiding catastrophic rework and an unlimited cycle of fixing flaws. With MBSE, the means of resolving issues is not unduly costly or time consuming. This is because measures exist to allow the systems engineering team to know when each design stage is complete, to see the direction for design advancement, to manage the impacts around resolving flaws, and to mitigate the consequences of uncovered design issues.

Requirement 2: The process must manage system complexity well.

As problems and their solutions become more complex, it is correspondingly more difficult to ensure that solutions are consistently and completely defined. Applying MBSE principles as an engineering and management process approach provides a powerful way of describing problems and their solutions. Because this description process is rigorous and complete, systems engineers are able to manage the problem complexity in a complete and disciplined manner.

As will be shown below, the integrated nature of the layered model is the key to managing complexity. It "automates" the tracking of the relationships and tracing paths that would otherwise need to be maintained and followed by hand. What would, absent the model's integrated structure, be an arduous task of ferreting out the information from multiple sources becomes the simple task of looking to the model, where the information has already been entered and maintained over time. Freed from the ministerial task of repetitively finding and recording information from a variety of repositories, the design team is enabled to leverage the power of the MBSE model to manage levels of complexity simply not possible with more fragmented and labor-driven approaches.

Requirement 3: The process must lead to effective solutions to a broad range of customer needs.

The MBSE approach provides a process that converges on a complete solution across a broad range of customer needs. MBSE is adaptable to the particular engineering problem type at hand, and the process advances the system design without performance disruption—even in the midst of unknowns. Engineering, operational, and social choices can be considered in advancing the design. In addition, the system can include the ability to have a nondisruptive means of accommodating on-going source requirement changes.

Requirement 4: The process must accommodate the three main problem classes (engineering unprecedented systems, reverse engineering, and middle-out engineering).

Because it is adaptable to the layers it finds in the problem definition, MBSE is useful in situations that require top-down engineering, reverse engineering, and middle-out engineering. In each instance, the engineering process begins where the needs exist and moves through the layers of the problem to the ultimate solution. Once the ultimate solution is reached, it is necessary to "prove" that the solution addresses the needs that drove the development project. MBSE makes this process manageable by proceeding through and documenting the development project in a way that is conducive to tracing the sufficiency of the system offered as a solution.

STRATA iterates through the primary concurrent systems engineering activities at each layer. The degree of effort within each domain varies according to the nature of the problem, the level being worked, and the boundary conditions affecting the design and process. But, as is illustrated in the gears shown in Figure 25, work in any domain influences the others. In a typical top-down design problem, the requirements domain tends to dominate the work at the first layer. In a reverse engineering (bottom-up) effort, however, the synthesis/architecture domain would tend to dominate the work at the first layer. Work advances across the domains (horizontally) in the layer while considering how the design should advance to the next layer. But note that the major thrust is to advance the work as indicated by the down arrows in Figure 25.

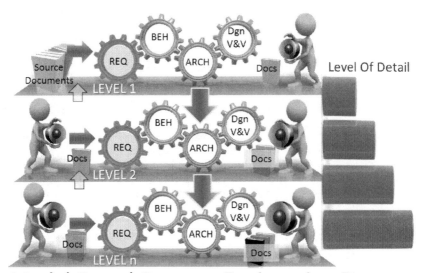

Model-Based Systems Engineering Process

Figure 25

Within each layer, the engineer must resolve ambiguities and make decisions on items that are "to be determined." In resolving such issues, the design decisions made at the previous adjacent layer may be brought into question and need to be revisited. The narrow arrows in Figure 25 indicate this adjustment in the process. This provides the engineering team the opportunity to make design corrections and refinements as a result of new insights, discoveries, or design decisions.

A key STRATA concept is that the engineering team should only have to iterate between adjacent layers. This allows the process to converge on a solution. Completeness and convergence are essential principles of STRATA. Therefore, a layer should be completed before advancing to the next layer (preserving completeness), and iteration back more than one layer becomes unnecessary. Should something be introduced into the problem (e.g., an external change) that necessitates revisiting earlier layers, the engineering team would be alerted to the creation of a major problem involving schedule and cost impacts resulting from having to reengineer the solution back through the layers to reach the point of process progress at which the issue arises.

Because it converges on the ultimate solution layer by layer, STRATA is truly a convergent process. This leads to it being "fail-safe." A truly convergent process always leads to a solution for any problem set that has a realizable solution, even given the constraints of cost and schedule. However, if circumstances (such as a redirection of resources) disrupt the process, the convergent process produces a draft system and other specifications at whatever process layer it encounters the disruption. This means that an abbreviated process does not result in a complete waste of the effort invested up to the time of interruption. The solution at that point is complete to the level of granularity represented in the last layer completed.

The system design process proceeds from a conceptual to a detailed description by moving downward with increasing detail or granularity from one layer to the next. Each of the iterations analyzes a layer within the system design process. Beginning with the basic, high-level user requirements serves to define broad system characteristics and objectives. Clarity is brought to these high-level expressions until the process has defined the system to a point of sufficient granularity to allow the system's physical implementation to begin.

Most systems engineering processes are not convergent precisely because they have difficulty managing process interdependencies and consistency across engineering domains under the constraints of satisfying cost and schedule. In those approaches, these factors tend to be treated disjointedly. By contrast, the STRATA approach enhances the project discipline by proceeding layer by layer across all domains, causing all factors to be addressed both systematically and systemically. The engineering team and project management, therefore, have better insight into the trade-offs needed to advance the design effort. This allows the domains to advance together, preserving completeness.

In addition, the engineering team can work consistently at the correct level within the design process. There are some necessary exceptions, such as risk reduction and addressing long-lead issues. For these issues, deep-dive studies are appropriate to ensure system success. Ultimately, however, the complete solution unfolds in increasing levels of detail and intermediary results are available for early review and validation.

MBSE Model and System Definition Language

Whether based on MBSE or not, all system design efforts develop their own language over time because of the necessity to communicate certain ideas clearly. The problem is that these ad hoc approaches rarely achieve the fundamental goals of clear communication. They take time to develop and, because the development is unplanned and ad hoc, this impedes system design progress while the language is established. MBSE avoids this situation because the language is developed in advance and eliminates confusion from the project's start.

The engineer uses a formal specification language to characterize the various design entities (requirements, functions, components, etc.) in a repository. Using this language and a repository allows the engineer to construct a systems "model." By capturing all the system information in the repository and correcting all errors, the engineer builds the repository to contain the system model from which the design team will produce the system, segment, and interface specifications.

Using the language constructs of the system definition language, the systems engineer can develop a model that illuminates both the physical architecture and the functional behavior of the system. The functions and the interactions of the system with the external entities enable the system to process the inputs into the outputs needed to satisfy the needs that drive the system. This discussion of the MBSE process now turns to the concepts of system function and behavior, and the development of system threads. Threads are sequences of behavior that trace defined paths through the system. Collectively these threads integrate to define the system behavior.

There are substantial benefits to maintaining system models throughout the system life cycle. In the development stage, the development team must characterize the system problem. This is an effort to define both the system context representation and the system boundary. The system boundary definition specifies the system of interest, as well as all external entities interfacing with the system. Thus, the development team identifies the system-level functional interfaces between the system and those external entities.

When working with top-down problems, the engineering team draws from the source documents (requirements, concept of operations, and other documentation) to identify what the system is and what it does at its most abstract level. This functional context model consists of the "root function" for the system and corresponding root functions for each of the external entities encountered. Thus, this diagram identifies all system-level stimuli and responses for the system.

The system boundary also identifies the limits of system development. The engineering team is responsible for everything within the boundary, handling all inputs, and developing all system outputs. In addition, they must manage all the system interfaces with external systems. Even though this high-level process appears overly simple, its purpose is to identify the principal inputs and outputs transiting the system boundary and to discover whether any external systems are missing. In many cases, this activity is not done at all or not done well.

With the originating requirements in hand, the engineer proceeds to layer 1.

Developing Layer 1 of Our Solution

Layer 1: Requirements

Work in the requirements domain begins at the highest level with relatively general statements from the system owners and stakeholders. These may take the form of a Concept of Operations (CONOPS), a Request for Proposal (RFP), internal customer documents, or all of these and more. From these descriptions of the system and its purpose, the requirements will begin to emerge.

As the design progresses through the layers, the requirements will be interpreted and refined into more and more particular statements. The more particular requirements are the "children" of their more general "parent" requirements.

Take, for example, the requirement in the sample problem that the system "accept information requests from certified customers." As the design proceeds, it becomes apparent that customers need to be able to make requests for images in several different media/formats. For instance, they need to be able to make a verbal request by phone. The system must be able to accept and process such a request. The initial requirement for accepting customer requests (the parent) is refined by the more specific requirement that the system "accept requests via telephone" (the child).

Layer 1: Functional Behavior

Once the requirements have been collected and analyzed at the level of the current layer, the next step in the design process is to design the functional behavior to implement the requirements. Like the other domains, the behavioral (logical) design advances in layers from the more abstract toward the more specific and complete functional representations. This is done concurrently with corresponding efforts in the requirements domain and the architecture/synthesis domain. The model interrelationships among these domains must be maintained to ensure completeness and convergence. This also minimizes the likelihood of having major rework because of decisions made at an inappropriate level in another domain.

Concurrently, an analysis of input/output transactions occurs in the requirements domain and in the architecture/synthesis domain to characterize broad classes of input/output transactions based upon the system boundary determinations that have been made. This identifies the stimulus-response characteristics of the system. The objective is to derive the possible threads to analyze in layer 2, which is a discovery process. System threads are generated for several reasons:

1. To make sure that every system input (stimulus) is properly addressed.
2. To assure that the system logic is fully developed.
3. To break the problem into solvable pieces.

In particular, the processing required for each class of system stimulus/input should identify a thread. This also enables the design team to identify system-level functions. Some of the questions addressed in this discovery process are:

1. Which inputs or input sequences lead to which outputs or output sequences?
2. Are there outputs or output sequences for which there is not an external input (stimulus)?
3. Are there inputs or input sequences for which there are not external outputs (responses)?

When the source documents lack sufficient information to determine these threads, the engineering team may suggest possible stimulus-response possibilities and engage the customer in approving these or identifying additional threads. The work products within the functional/behavior domain are sometimes collectively referred to as the "functional architecture" of the system.

Layer 1: Architecture

Source documents serve to define the top-level architecture and functional context. Its purpose is to assure the identification of all the externals and environmental entities. Thereafter, the context aids in discovering all the principal input and output classes transiting the system boundary. This identifies the primary system-level interfaces. Each layer advances the completeness of the design and influences the work in the other domains. Conversely, the work in the other domains influences the physicality of each layer. Handling this interdependence is foundational for finding balance in the system design. MBSE does this in a disciplined, orderly fashion.

Just as the functional/behavior domain is a part of each layer of the STRATA, so is the architecture/synthesis domain. Developed in concert with the analytical work on the requirements domain, it is influenced by allocation decisions made in the behavioral domain. The objectives of the architectural model are determined by the current level of the system model. For example, layer 1 architectural model objectives are to identify the system and the external counterparts with which it must interact. Accordingly, layer 2 architectural model objectives seek to discover physical partitioning strategies for the system. Subsequent layer-by-layer objectives investigate, refine, and evaluate the strengths and weaknesses of each partitioning strategy employed. They seek to maintain the relationships among all the other domains—preserving each layer's boundary conditions and maintaining the evidentiary path for system acceptance.

Architectural Language

There is a clear need for a language to express the system's architectural design. The language of architecture/synthesis is both graphical and textual. The graphical language used to support architecture/synthesis in model-based systems engineering is abstract and represents the hierarchical structure of the physical design and the interface relationships among internal and external components. (These are expressed in physical hierarchy diagrams and physical block diagrams.)

Just as in the behavioral domain, it is preferable for these architecture/synthesis graphical views to be generated from the model. This avoids having to exhaustively review all diagrams to determine what model changes affect current views, making long-term model support less labor intensive and, therefore, more manageable. Showing physical structure graphically requires that the graphical language (notation) possesses certain characteristics. These are similar to behavior characteristics. The language must:

1. Be understandable (allowing for information hiding/abstraction)
2. Preserve physical hierarchy (supporting decomposition and aggregation)
3. Support behavior (decomposition, aggregation, and allocation)
4. Support behavior executability

Therefore, in the system definition language (SDL), components are decomposable; that is, they have parent-child relationships. The SDL must provide for the connectivity of links and interfaces to express how components physically relate while under the condition of preserving behavior (observables and exit criteria).

Matching physicality with behavior is an important property of the system definition language. Physical decomposition must preserve behavior under decomposition and aggregation. This is necessary to allow the logical design and the physical design to be consistent through each design layer. Modeling inconsistencies within each layer and between layers indicate system design flaws. These flaws may well be critical and, consequently, the system design itself may be seriously flawed. A process such as MBSE that enables system designers to easily identify these inconsistencies is, therefore, critical to producing a successful design.

Behavior preservation under allocation is necessary for maintaining consistency with the integrated logical model. The process should result in the same system behavior when the partitioned behavior is allocated to system components as was presented in the behavioral model itself. A simulation of the integrated logical model and a simulation of the allocated model should give identical results.

Because of the need for process documentation, all observables need to be preserved under both decomposition and aggregation. Just as in the behavioral domain, inputs and outputs, input and output sequencing, number of and conditions for exits, and performance must all be preserved. Performance characteristics under decomposition need to be carefully managed. Performance measures may change at different levels of decomposition and performance characteristics—especially time performance values—may need decomposition as well.

The preservation of system behavior across decomposition is much less likely to occur when the systems engineering approach tends to focus on components/objects first, because one implicitly rather than explicitly allocates functionality to the components regardless of the true needs of the system or the system's users. Implicit behavioral allocation makes it likely that physical decomposition may not always preserve the conditions necessary for logical and functional consistency throughout the system design. This is the price of the loss of the "systems view" that so often afflicts component-driven engineering efforts.

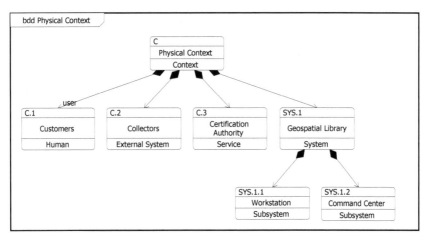

Figure 26

A representative physical hierarchy diagram is shown in Figure 26 along with the physical block diagram variations of Figures 27 and 28. The physical hierarchy diagram shows, organizationally, the system's physical architecture from the most abstract to the concrete representational aspects. It is the concrete, or lower-level, physical components that are actually specified and built. In Figure 27, the system and its context is revealed at layer 1 of the hierarchy. Layer 2, in this example, reveals the two concrete components that the system comprises (in this case, the Geospatial Library).

A Primer for Model-Based Systems Engineering

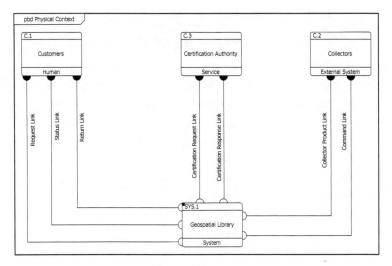

Figure 27

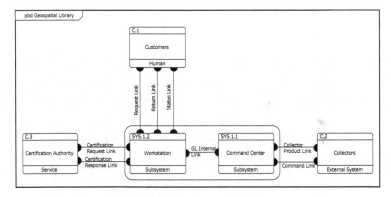

Figure 28

Correspondingly, the physical block diagrams present the physical interrelationships among the components. Figure 27 reveals the interconnections with the entities external to the system being developed. Figure 28 provides a view of the interrelationships among the external entities and between the system entities.

Architecture Design at Layer 1

In the development stage, the system representation is characterized and presented to the development team. The fundamental effort at layer 1 is to define and express the system's context and system boundary. The system's boundary definition identifies the system of interest and all external entities interfacing with it.

The systems engineering team uses the source documents (requirements, concept of operations, and other documentation) to identify what the system is and what it does at its most abstract level. The physical context view, which is analogous to the functional context view, consists of the system and the external entities encountered as well as the stimuli and responses crossing the system boundary.

The system boundary identifies the limits and focus of our system development. The engineering team is responsible for designing everything within the boundary, handling all inputs, and developing all system outputs. Even though this system context diagram appears overly simple, its purpose is to identify the principal inputs and outputs transiting the system boundary and to discover whether any external systems are missing; yet in many cases, this activity is not done at all or not done well.

From an architectural point of view, the task at layer 1 is to use the context diagram to show both the physical entities of the system as well as those external entities that will interact with the system. In doing this, the top-level architecture follows the top-level behavior.

Care should be taken with the definition of the system boundary. The boundary should be drawn to cast the net neither too broadly nor too tightly. An improperly selected system boundary either adds unnecessary entities to the system design (too broad) or, even more detrimentally, excludes necessary entities from the system design (too tight). Excluding necessary entities adds more ambiguity to the design effort and creates more difficulty in resolving those ambiguities. An improperly selected system boundary affects the tasks associated with the requirements, behavior, and architecture/synthesis domains.

Proceeding with Layer 2

The layer 2 objective for system behavior is to develop the integrated logical view of the system. The integrated logical view is the integration of the system threads for idealized system behavior; that is, the system behavior without addressing faults, errors, resource management, and so on. The approach is to model the behavior of the threads identified during the work on layer 1. After completing these threads, the systems engineering team builds the integrated model. The integrated model incorporates the functional aspects and insights from each thread into the integrated model.

The goal in establishing the system behavior is to provide a specification of what the system must do to meet the functional requirements without inferring or assuming any particular technical solution (the physical structure and make-up of the system). Maintaining this separation (between the behavior and a particular solution) requires a surprising amount of discipline. Though this separation may not always need to be total and absolute, care must be taken that assumptions about the physical architecture are not made so prematurely that it creates artificial constraints on the system design.

Thread Development

Generating a thread begins with selecting one of the classes of system input (this input is a stimulus for the system). The next step is to create the sequence of system functions necessary to respond completely to that input. This sequence normally results in the creation of one or more system outputs. Some threads may not terminate as a system output, but may enter data in an internal database or change the state of the system.

At the same time, the appropriate source functional requirements are associated with these thread functions. When the thread development is complete, all the functional requirements should have been addressed in one or more of the threads. If that is not the case, then something is missing from an existing thread or another thread needs to be developed.

As an example, consider a thread from our example system providing products to users from an existing inventory in response to customer orders. Figure 29 shows an enhanced functional flow block diagram of this thread. This could also be shown just as easily in an activity diagram like the one in Figure 30.

Figure 29 postulates a sequence of functions that responds to a customer request and is limited to the case where the product already is present in the system's inventory (thread 1, labeled t1 in the diagrams). The thread's flow is from left to right, as shown by the directional arrows on the branch. The stimulus for this thread is the information request, which is placed by the customer (t1.1 Make Information Request). The system performs the next three functions, and the customer performs the last function (t1.2 Accept Products).

A Primer for Model-Based Systems Engineering

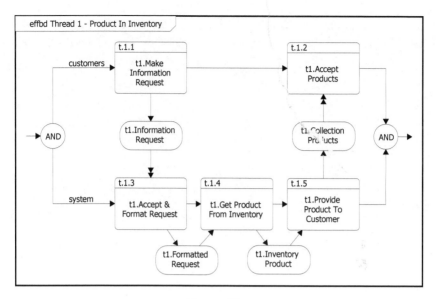

Figure 29

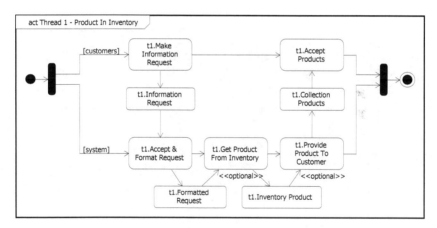

Figure 30

A second thread (thread 2, labeled t2 in Figure 31 on page 88) considers the case created when the product requested is not in inventory. In that instance, it is necessary to identify the needed product and procure that product from an external system.

A Primer for Model-Based Systems Engineering

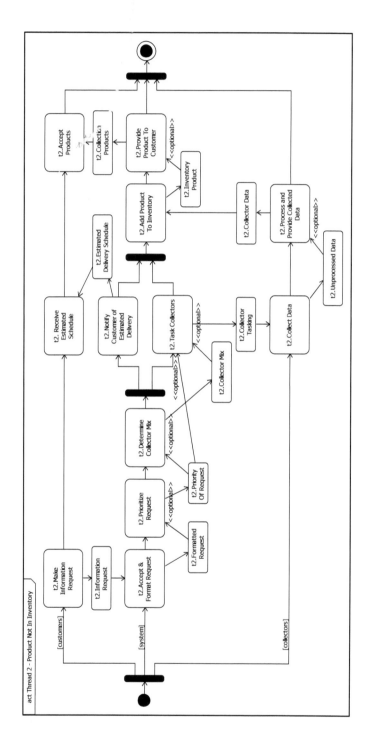

Figure 31

Figure 31 is more complex and shows logic that requires the functional involvement of the system with two external entities. The functions performed by the customer are on the top branch and include making the initial request, receiving the estimate regarding the anticipated delivery date for the customer's order of products, and accepting the products on delivery.

The functions performed by the collectors are on the bottom branch and include collecting the data in response to the tasking from the system, as well as processing and providing the collected images back to the system for placement into the system inventory. The system performs the remaining functions.

These two threads cover the two possibilities that result when a customer makes a request. Once the system accepts the request, it either does or does not have the image in inventory. If it does, the functional behavior follows thread one and the image is retrieved and provided to the customer. That case is covered by Thread 1. If the system inventory does not include the requested image, the system must task a collector to procure it, add the image to the inventory, and provide the image to the customer. That case is represented by Thread 2. Together, they represent the operation of the Geospatial Library.

After generating the system threads for every class of system input, it is necessary to check to see whether every system output identified in either the system context diagram or the system requirements has been addressed. If not, there must be threads that have not yet been identified or outputs that are the result of internal triggers (e.g., a requirement for periodic system health and status data). These internal triggers must be identified and threads generated for each case. These threads form the basis for creating the integrated logical model and should be saved for later reference to help, for example, formulate system test threads.

Once the systems engineering team completes the system threads, they develop the integrated logic to make the logical model more compact and understandable, and to take advantage of commonalities among the threads. At this point, the logical model represents the system's idealized behavior. Subsequent work adds complexity to address error conditions, resource management, failure recovery, security, and other needs. Each layer advances the completeness of the design and influences the work in the other domains. In the same way, the work in the other domains influences the functionality of each layer. Designing with this in mind is fundamental to finding balance in the system design.

Thread Integration

Once the set of system threads is complete, they must be integrated and any commonality of functionality among them accounted for. Systems engineers must also account for the logical interaction and logical control for that combination of threads.

It is tempting to integrate these threads by keeping them in their original form and representing them in a multibranched parallel construct. This strategy actually makes it difficult to define and/or understand system function interactions as an integrated whole. Rather than simply aggregating the threads together, it is helpful to integrate their behaviors into a synthetic whole.

The objective is to integrate the threads in a manner that minimizes the size and complexity of the final integration. This is done by minimizing the duplication of functions and logic streams. The set of system threads identifies how the system responds to each input independent of all other inputs. This is an effective strategy for managing complexity. However, the individual threads do not account for commonality of functional process, nor do they account for thread interactions.

One common integration problem is that different individuals generate the different threads. That means that similar functions may be described as having different names, different boundaries, and different inputs and outputs. The resolution of these differences must be a part of integrating the threads into a single, common architecture.

While some differences are differences in name only, others are functionally real. Functions with differing inputs and outputs are actually different functions. Likewise, functions with different exit conditions are different functions. This must be taken into account when defining the unique system functions during the integration of the threads.

The integration process defines the means of control over the integrated whole. This activity may prove to be a challenge, but it is essential in order to gain an efficient and understandable set of system logic. It is usually easier to integrate the threads by looking for common entities. Examples might be inputs from the same external system, inputs through the same input channel, or inputs requiring the same initial functional processing. This integration activity is a creative challenge dependent on the insight gained during thread development, and there is no standard formula or process for it.

The resulting integrated behavior diagram contains the threads involved in the integration. Each of them can be seen as an identifiable path. Figure 32 is an enhanced function flow block diagram (EFFBD) resulting from the integration of the individual threads into a single integrated behavior diagram. This depicts the functional architecture of the system. It consists of the structured sequences and logic of system functions, including the inputs, outputs, and triggers which relate to the functions. When integrated with the enhanced EFFBD for all of the external systems, the resulting logical model becomes executable.

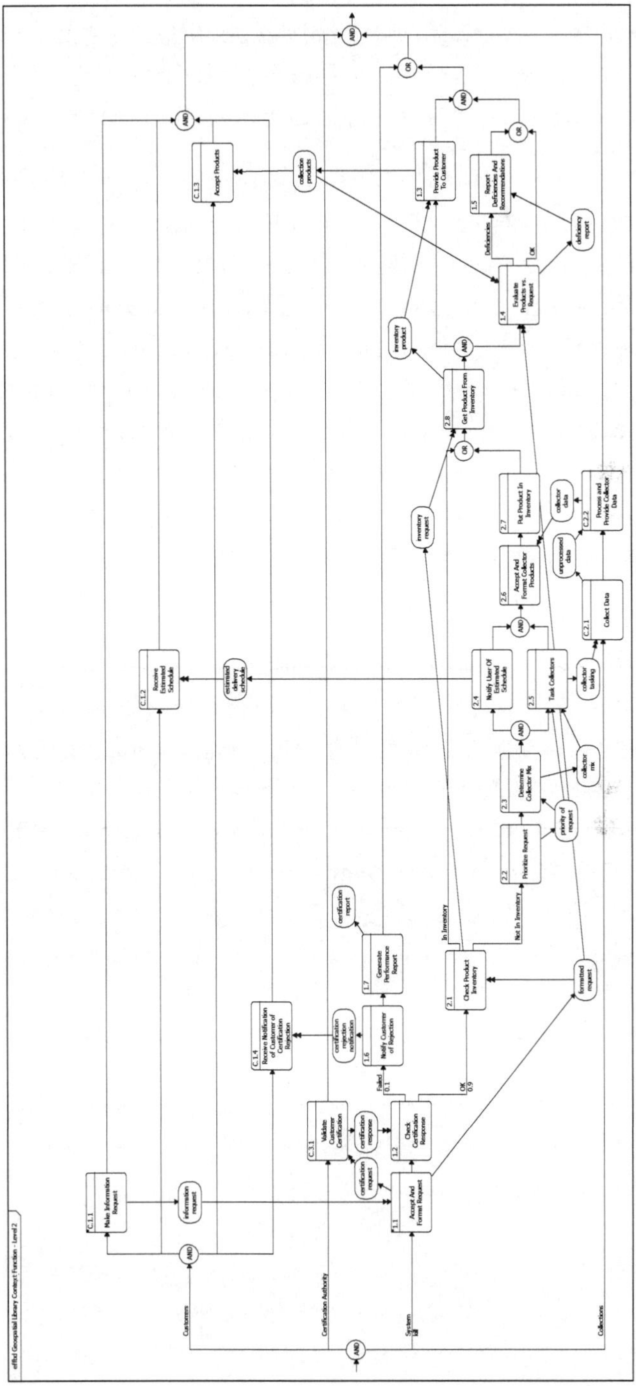

A Primer for Model-Based Systems Engineering

Figure 32

As the functional design progresses, additional detail is added to address other functional support needs. These needs are error handling, fault recovery, input overload, resource management, and the additional control logic to manage the additional functional logic.

At each layer of the model, satisfaction of the layer's completion criteria has to occur before advancing to the next layer. Part of those completion criteria involves assuring that the functional model's boundary conditions with the other domains are met. That is, source functional requirements are traceable to the derived lower-level functions, which in themselves are derived functional requirements. In addition, these same lower-level functions must be properly allocated to the components of the physical architecture.

Architecture Design at Layer 2

The architecture/synthesis objectives for layer 2 are to develop and apply an effective partitioning strategy for the physical components in response to the behavior model. Partitioning requires finding a balance between functional groups and the components to which these functional groups are assigned or allocated. System partitioning serves to find a balanced set of physical components that are relatively easy to construct, and integrates them into a useful and usable system for the system's stakeholders.

The work in layer 2's behavioral domain, the integrated logic model, is a reasonable starting point for this process, because at this point the integrated logic model usually possesses more structure than the current physical model coming out of layer 1. If there are any physical architecture constraints, such as using existing components or contractual required configuration items, then they affect the process.

In those instances, the architectural constraints become fixed points in the physical hierarchy and must influence the integrated logic to assure that the behavior of the affected components is easily recognizable and allocatable within the integrated logic model. The allocation is an iterative process affecting the integrated logic in an effort to find a balanced set of functions to allocate to a corresponding set of components. Work in the behavioral domain influences the system's component set and vice versa until a balance is found.

Effective partitioning requires criteria for evaluating the various partitioning strategies and the results of applying those strategies. Among the primary potential partitioning strategy evaluation criteria are interface complexity, testing complexity, and performance partitioning among subordinate components. Secondary criteria include technology risk, future performance requirements, and future technology insertion. It is not necessary to use all these criteria, but there should be defined criteria. Once these factors have been reasonably met, the physical architecture is reviewed by the specialty engineering teams to assess the feasibility of building a system based upon the proposed system partition.

Architecture Design at Layer *N*

As the layered design progresses, additional functional detail is added to the behavioral model. Other functional support needs, such as error handling, fault recovery, input overload, resource management, and the additional control logic to manage the additional functional logic, are addressed. As a consequence, the architecture/synthesis effort needs to readdress the physical partitioning of the system to accommodate the added functionality and its refinements.

The same evaluation criteria are applied to each alternative partitioning approach to reveal the better physical architecture. Alternative architecture partitions are brought forward only when the architecture cannot be clearly rejected based on the acceptance criteria.

At each layer of the model, satisfaction of the layer's completion criteria must occur before advancing to the next layer. Some completion criteria assure that the architecture/synthesis partition's boundary conditions meet the constraints of the other domains. That means that source functional requirements are traceable to the derived lower-level functions, which are themselves derived from functional requirements.

In turn, these same lower-level functions must be properly allocated to the components of the physical architecture. Each evidentiary artifact is revisited to capture the current design factors contributing to satisfying the system's acceptance criteria. If necessary, additional evidentiary artifacts are generated or captured. This assures that the design process is completed.

The physical design advances in layers from the more abstract toward the more specific component representations. This occurs concurrently (layer-by-layer) with corresponding efforts in the requirements domain and the behavioral domain. The model interrelationships among these domains must be maintained to ensure completeness and convergence. Doing so also minimizes the likelihood of having major rework because of decisions made at too low a level or in another domain.

As work progresses from layer to layer, trial behavioral allocations are evaluated and architectural design decisions made based on the results. By proceeding from layer to layer in a logical fashion, STRATA produces an increasingly more detailed model of the system, with a sound behavioral allocation at each layer. Because the model is constructed with fully documented relationships, each layer produces a model which can be tested and simulated in order to prove the integrity of the allocation against the requirements.

Verification and Validation

Verification and validation is not a single, culminating event leading to system acceptance. There are a number of intermediate steps occurring across the layers of the model. As the system design progresses, evidence is gathered from the engineering activities within each layer. That evidence becomes the trail needed for constructing the argument for the ultimate system acceptance as well as for verifying that the work meets the requirements and objectives of the system along the way.

For every layer of the model, at least one design review occurs to gain consensus that the layer's model is complete and consistent to that point. This validates that layer's model and design and allows the systems engineering team to move on to develop the next layer.

The relative ease or difficulty in gaining system acceptance depends heavily upon the system integrity maintained throughout development. In this sense, system "integrity" means that all the intermediate and final work items are traceable throughout the process, and decisions made along the way are rational and defensible (with respect both to engineering and management processes).

Some of the intermediate work items become evidence of the system integrity in that they support the conclusion that the system satisfies all the expressed needs. Thus, the evidence needed to complete a formal verification and validation process is developed and preserved throughout the development process. Ultimately, the customer should be confident that the system possesses no fatal flaws or exploitable vulnerabilities and that it has supportable components. In short, the system must be usable and useful for the purposes for which it was intended. The design Verification and Validation domain is where the integrity assurance activities occur.

As we have seen, each layer of the STRATA model involves its own allocation activities. These establish how the functional/behavioral, architecture/synthesis, and requirements domains interrelate. Allocation is the activity that apportions entities in one domain to entities in another domain.

Formal verification and validation processes demonstrate that the delivered system meets the customer's needs and satisfies the design contract. The basis of formal verification and validation is demonstrating that the delivered system satisfies the needs driving the project, is useful, is usable, and answers the agreed-upon requirements.

Formal Verification and Validation

In traditional systems engineering approaches, requirements reviews most often occur without adequate allocation to the physical or logical representations. Because the model-based approach addresses the allocation systematically, it leads to a better-grounded method for validating the system design.

One aspect of the review is the validation of the requirements. Validating requirements ensures that the set of requirements is correct, complete, consistent, and traced appropriately to model entities. As the layers of the model develop, requirements are added through design decisions, derivation, and layer-specific requirements found in the source documents. The validation review progressively encompasses more requirements at each layer and builds upon the conclusions of previous reviews. This progressive process leads to both a better understanding of and greater confidence in the validation process.

Simulating the logical model as a part of the process enables evaluation of "dynamic" model consistency. Executing the model through simulation uncovers dynamic flaws that are resolved through correcting and refining the logical model and, in turn, results in requirement changes. Capturing these requirement changes and the reasons for these changes maintains the model's integrity.

Verification shows that all the requirements produced through the system design process are indeed satisfied within the physical instantiation of the system and its components. Proper verification depends on the trail of artifacts developed throughout the design process as well as those resulting from the operational simulations of the system. The discipline, consistency, and convergence of the MBSE processes provide a trail for the verification and validation processes to follow. This justifies a high level of confidence in the decision to accept the system.

System Acceptance: Requirements Verification

Requirements verification requires a strategy for showing that the implementation of the design achieves the design's objectives and meets the acceptance criteria. The strategies for verifying constraint, functional, and performance requirements are generally different. For example, the strategy for showing the satisfaction of maintainability requirements differs from that for showing the satisfaction of weight requirements. The requirements for executing those strategies are called verification requirements. These requirements shape the system acceptance testing, which serves as the means for verifying constraint, functional, and performance requirements.

Ideally, there should be a minimal set of verification requirements. The smaller the number of verification requirements, the fewer tests, inspections, analyses, and other verification activities necessary to show that the implemented system does what the design claims it should do. The system model's leaf-level requirements trace to verification requirements, and the verification requirements trace to a verification method appropriate to prove that each lower-level requirement is satisfied by the implemented system. Verification requirement methods may also trace to tests, test plans, and so on. For functional requirements, the integrated logical model and even threads assist in deriving a suitable set of test cases. Capturing the test results in the system model layer by layer helps build the evidence needed for ultimate system acceptance.

Defects uncovered during testing need resolution. The verification method which discovered the defect points to the verification requirement. From the verification requirements, tracing back to a set of lower-level requirements helps identify what system entities contribute to the uncovered defect. Analyzing these areas leads to finding the defect's cause and its resolution. After correction, repeated testing reveals whether the system deficiency is resolved.

It is easy to see the value of an integrated model where all of these linkages are maintained within the model itself. That integration both eases the workload in making the accurate and complete tracings required across the domains and improves the quality of the result. The integration and linkages provide an assurance of completeness for the analysis and testing.

System Acceptance: Validation

Operational testing is the typical approach for validating the system for acceptance. The objective is to demonstrate that the system is the "right" solution—one that is usable, useful, and fulfills the customer's needs. The validation process builds on the verification tests supporting the system acceptance decision. Operational testing invokes various quality criteria along with measures of effectiveness and performance to assess how well the system fulfills its purpose. Testing approaches may include such measures as stress testing, load testing, and failure modes. As in system-acceptance verification testing, the verification requirements and methods continue to show traceability, making them evidence for acceptance when such tests are satisfactorily concluded.

Verification and validation of requirements, system design, and system development are continuing processes in model-based systems engineering. Verification and validation take place at each layer of the model. Artifacts generated under the aegis of verification and validation become part of the documentation that leads to final system sign-off by the customer.

In its system definition language, tools, and processes, model-based systems engineering incorporates the structure to develop systems that fulfill all the objectives of the customer. The MBSE processes eliminate inconsistencies, errors, and omissions early in the design and development stages. The chain of artifacts that will ultimately make the case for system acceptance is generated link by link at each layer. By reducing the likelihood of failing to uncover catastrophic defects until late in the development process, MBSE reduces inherent program risks. This is a direct result of the consistent and convergent nature of the MBSE processes.

The convergent nature of MBSE naturally supports the needs of the verification and validation processes. Having used MBSE affords the assurance, through verification and validation, that the system does indeed fulfill the purposes for which it was designed and, in so doing, completely satisfies all of the system requirements.

SUMMARY

There is not unanimity around the definition of model-based systems engineering in the marketplace, and most uses of the term model-based systems engineering are not as broad as what is addressed here. By intentionally adopting a "broad" definition of a system, we have tried to show that the approach can be used across the widest possible spectrum of systems to be analyzed or constructed. Clearly, from the definitions and discussion, the aim of a system model is to be able to analyze and gain insight into real systems, whether human-made or not. With human-made or "engineered" systems, the aim is to find a realizable solution to a stated need using effective engineering and management processes.

The same iterative approach that is used to define and design a new system (the top-down engineering approach) can be used to analyze and improve an existing system (middle-out or reverse engineering). In those cases, the existing system becomes a set of requirements and constraints for the design.

In any case, applying an iterative, convergent (layer-by-layer) systems engineering process reduces ambiguity by resolving open and uncovered issues and mitigating risks. Systems engineers and stakeholders collaborate with other team members to make decisions that advance the design to completion. When the design is complete, validation and verification can take the form of "walking through the design," verifying that all the requirements are valid and can be verified, all the functions are present that are necessary to meet the requirements, all appropriate analytical and simulation activities have been performed, and all components needed to perform the functions at this level are defined. In other words, the engineers and stakeholders can verify that the layer-by-layer process has converged on an engineering solution that satisfies all of the requirements for the system being designed.

This represents a huge advantage over the more traditional document-driven approaches. Where they are slow to respond and do not necessarily converge on a solution, layered, iterative MBSE offers an advancing design at every stage. By maintaining the systems view throughout the problem-solving process, MBSE also offers a significant advantage over the more agile approaches to engineering design. With a foot in both worlds (disciplined system view and responsive design), MBSE positions the design team for success.

AFTERWORD

The MBSE approach offers the system owner significant advantages in designing a new system or improving an existing one. Because the design proceeds in an orderly, logical and convergent manner, it reaches a solution that answers the owners' needs with a high degree of confidence. Because it proceeds to "peel the onion" layer by layer, it offers a complete solution, consistent with each layer's constraints, at any point in the process. If resources or other contingencies interrupt or halt the development program, the solution is usable and the resources expended to that point are not completely lost. Because the model uses a clear, unambiguous language to describe the problem space and the solution set, the design process can use the expertise of a diverse set of contributors without the typical problems of confused and inefficient communications marring the outcome. In short, the system stakeholders can have confidence that the MBSE design process will converge on a solution that is useful and usable in meeting their needs.

AUTHORS

David Long founded Vitech Corporation in 1992 to develop and commercialize CORE®, a leading systems engineering software environment used globally. He continues to lead Vitech in delivering innovative solutions and empowering organizations to develop and deploy next-generation systems.

For over twenty years, David has focused on enabling, applying, and advancing model-based systems engineering (MBSE) to help transform the state of the systems engineering practice. He has played a key technical and management role in refining and extending MBSE to expand the analysis and communication toolkit available to systems practitioners. David is a frequent presenter at industry events worldwide delivering keynotes and tutorials spanning introductory systems engineering, the advanced application of MBSE, and the future of systems engineering.

Zane Scott manages Vitech's Professional Services and Training organization. For the past twenty-five years, Zane has built a skill set which enables him to provide insight and guidance to individuals and companies as they improve organizational processes and methodologies. Zane has also taught systems engineering methodology in the practical process context in a variety of settings.

Zane brings a unique perspective to Vitech and its customers. With a professional background in the litigation field, Zane is also a trained negotiator, labor management facilitator and mediator. He has practiced tactical negotiation and interventional mediation and taught communications, conflict management and leadership skills at both the university and the professional level. Before joining Vitech, Zane worked as a senior consultant and process analyst assisting government and industry clients in implementing and managing organizational change.

Made in the USA
San Bernardino, CA
23 October 2015